"It's only a matter of time before a really large space object hits the Earth. Asteroids and comets have created disaster many times in the Earth's past, including the impact that wiped out the dinosaurs, and the vestiges of their havoc are still apparent. Gordon Dillow's enthralling discussion unlocks the secrets of how and why these objects jeopardize the planet and what thousands of people around the globe are doing to detect and defend against them. *Fire in the Sky* is nonfiction that reads like a great adventure novel, even as it points toward a hopeful future for humanity."

—Roger D. Launius, former Chief Historian of NASA and
author of *The Smithsonian History of Space Exploration*

"With this book, Dillow brings to readers' attention a serious threat to our planet. It's entirely possible that good science and R & D will be enough to avert catastrophe, but the mere fact that humanity could find a way to contend with what's hurtling toward us makes the threat no less real. What seems clear from Dillow's tour of the front lines of asteroid defense is that we as a nation must find the will—and the funds—to build the necessary tracking and stopping tools. Mother Earth is depending on us."

—David Livingston, founder and host of *The Space Show*

"Lucid and engaging . . . Dillow stresses that the threat is real, that the Earth is routinely hit by objects from outer space, and that it is certain that sometime in the future—maybe in the coming decades, maybe millions of years from now unless mitigating actions are taken—one of those objects will be large enough to cause catastrophic damage. . . . A convincing case for the need to pay more attention to planetary defense."

—John M. Logsdon, founder of the Space Policy Institute at
George Washington University, NASA advisor, and editor of
The Penguin Book of Outer Space Exploration

"Informative, timely, and entertaining . . . a great read! Dillow's treatment is never dull—often humorous—and provides accurate information about the resource potential of Near-Earth asteroids, their impact threats to Earth, and the ongoing activities to mitigate these threats."

—Donald K. Yeomans, author of *Near-Earth Objects: Finding
Them Before They Find Us* and former manager of the Jet
Propulsion Laboratory's Near-Earth Object Program Office

COSMIC COLLISIONS,

KILLER ASTEROIDS,

AND THE RACE

TO DEFEND EARTH

FIRE

– IN THE –

SKY

GORDON L. DILLOW

SCRIBNER

NEW YORK   LONDON   TORONTO   SYDNEY   NEW DELHI

Scribner

An Imprint of Simon & Schuster, Inc.
1230 Avenue of the Americas
New York, NY 10020

First Scribner hardcover edition June 2019

SCRIBNER and design are registered trademarks of The Gale Group, Inc.,
used under license by Simon & Schuster, Inc., the publisher of this work.

For information about special discounts for bulk purchases,
please contact Simon & Schuster Special Sales at 1-866-506-1949
or business@simonandschuster.com.

The Simon & Schuster Speakers Bureau can bring authors to
your live event. For more information or to book an event,
contact the Simon & Schuster Speakers Bureau at 1-866-248-3049
or visit our website at www.simonspeakers.com.

Interior design by Kyle Kabel

Manufactured in the United States of America

1   3   5   7   9   10   8   6   4   2

Library of Congress Cataloging-in-Publication Data is available.

ISBN 978-1-5011-8774-2
ISBN 978-1-5011-8776-6 (ebook)

To my mother and father
Louise Blackwell Dillow and Troy O. Dillow

CONTENTS

INTRODUCTION   I

Chapter 1   IMPACT!   7

Chapter 2   MIRACULOUS APPARITIONS IN THE AYRE   23

Chapter 3   ASTEROID MINERS   45

Chapter 4   STAR WOUNDS   75

Chapter 5   T-REX WITH A STRING OF PEARLS   105

Chapter 6   ASTEROID HUNTERS   139

Chapter 7   PLANETARY DEFENSE   163

Chapter 8   ASTEROID KILLERS   179

Chapter 9   ASTEROID WARS   203

EPILOGUE   229

ACKNOWLEDGMENTS   235

CHAPTER NOTES, SOURCES, AND RELEVANT FUN FACTS   239

INDEX   269

INTRODUCTION

Just before 4 a.m. on June 2, 2016, I was enjoying a pre-dawn cup of coffee on the back porch of my home in Arizona. And suddenly it was as if the sky was on fire.

It began with a glow that spread across the steep hillside behind my house, a kind of angry molten light, all red and orange and amber, the color of lava. The glow grew more and more intense, lighting up the mesquite trees and saguaro cactuses and casting their long, tortured shadows on the ground. It looked like something from the netherworld, like high noon in hell. Then, seconds later, way up in the sky off to the northeast, there was a blinding, thermonuclear-style flash, a burst of white light almost as bright as the sun, followed later by a sound like distant thunder that set my dogs to howling.

As the light faded and the night returned I stood there, transfixed, not quite believing it. As a soldier and journalist I've traveled the world for decades, and experienced all manner of cataclysmic natural and unnatural events—wars and riots and all manner of mayhem, typhoons, tornadoes, major earthquakes, even (from a distance) the deadly volcanic explosion of Mount St. Helens in 1980. But this was easily the most astonishing natural event I had ever seen. It occurred to me that this was how the world will end: with a flash of light, and a roar like God's own artillery, and then—darkness.

Of course, the world didn't end on June 2, 2016. And in the age of 24-hour news and social media, an explanation for this amazing phenomenon was soon forthcoming. Apparently a small asteroid, a rocky piece of space debris only about six feet across, had wandered into the Earth's path and exploded in the sky some fifteen miles above Arizona's White Mountains. The asteroid's fiery passage through the atmosphere and subsequent brilliant explosion lighted up thousands of square miles of ground and startled observers as far away as Texas. Despite the early hour, the event had been captured on scores of dashcams and smart phones, so although no one was killed or even slightly injured, the Arizona "fireball" was a major story not only on local TV but on the national newscasts as well. And as I watched and read the news reports, a couple of things quickly caught my attention.

One of them was the almost unbelievable power of the explosion. Soon after the event, the National Aeronautics and Space Administration (NASA) reported that the little asteroid had burst apart in the atmosphere with the energy equivalent of half a kiloton of exploding TNT—that is, a million pounds of TNT. To put that into perspective, the U.S. military's most powerful non-nuclear bomb is the GBU-43/B Massive Ordnance Air Blast (MOAB) bomb—the so-called Mother of All Bombs—which has a blast yield of a mere 22,000 pounds of TNT. That little asteroid made a MOAB look like a firecracker.

And there was another thing about that fireball in the Arizona sky that surprised me. That was the fact that no one had seen it coming. Sure, the streaking fireball and its subsequent explosion had been picked up by some U.S. spy satellites, by ground-based Doppler weather radars and by NASA's All-sky Fireball Network, a national network of cameras set up to record events like this one.

But that only happened *after* the asteroid entered the atmosphere, not before. That space rock hurtling out of the clear black sky was a complete surprise to everybody.

As often happens, the more I learned, the more questions I had. Such as, how in the world could a piece of rock the size of a La-Z-Boy recliner pack an explosive wallop almost fifty times greater than the most powerful conventional bomb in the U.S. military arsenal? Why in this era of satellites and space-mounted telescopes and world-spanning radar networks did no one spot the thing before it arrived? How often do events like this happen? What's the history behind our understanding of Earth-colliding asteroids and comets?

Those are some of the questions this book will try to answer. And it will also take a look at the Big Questions: What are the chances that a much larger chunk of space rock, something hundreds of yards or even miles wide, will find itself on a collision course with Earth? And if that happens, what if anything will we be able to do about it?

Actually, I can answer that first Big Question right now. The chances that an asteroid or comet of potentially catastrophic size will come hurtling toward Earth are exactly 1-in-1. It's 100 percent, a sure thing, a lead-pipe cinch. The only variable is when. It could be a hundred thousand years from now—or it could be next Tuesday. True, the U.S. government's comforting official position is that for at least the next century or so, there is no significant risk of Earth being struck by any large asteroid or comet *that we currently know about*. But given the fact that our Solar System is home to billions of asteroids and comets that we *don't* know about, that's a pretty significant qualifying clause. It's a loophole you could drive an asteroid through.

As for what we could do or would do if we spotted a potentially damaging space body headed our way—well, that remains to be seen.

Should we be worried about this? Should we as individuals be concerned about the threat of cosmic impacts? Yes, but not to the point where we toss and turn all night over it—at least not yet. But we should understand that the threat is real; it's science fact, not science fiction. And we should expect—even insist—that somebody pays attention to the problem. Fortunately, as we'll see, a relative handful of men and women are dedicating their professional lives to assessing and planning for these threats from space. But one of the themes of this book is that there should be more resources committed to the impact problem. To do otherwise seems foolishly short-sighted.

I have to confess that when I began this book I knew next to nothing about asteroids or comets or other things astronomical. I don't say that with any perverse, anti-brainiac sense of pride, the way some people boast about being terrible at math. It truly made me regret the shocking shortage of science classes in my college transcripts. Like most people, I had heard about the dinosaur-exterminating giant space body that hit Earth sixty-five million years ago, and I vaguely recalled the silly mnemonic for the order of the planets: *My Very Energetic Mother Just Served Us Nine Pizzas*, for Mercury, Venus, Earth, Mars, Jupiter, Saturn, Uranus, Neptune, and Pluto—although even that was outdated, since poor little Pluto was kicked out of the planet community years ago and reduced to mere dwarf planet status. And of course I had seen some of the giant-space-rock-on-a-collision-course-with-Earth movies that Hollywood has been churning out for decades—although none of them offered much in the way of expanding my scientific knowledge, or anyone else's.

So it was a little daunting to suddenly find myself immersed in a world of astronomical units and albedos and arguments of perihelion and the kinetic energy of hypervelocity impactors—which, as I'm sure everyone already knows, is equal to one-half the mass times the velocity squared, or $E_k = \frac{1}{2}mv^2$. I had a lot of catching up to do. There were hundreds of books and scholarly papers to read, countless hours spent perusing past and present articles in *Sky & Telescope* and Space.com and EarthSky.org and the Planetary Society website, and dozens of interviews with astronomers and planetary scientists and other people who are a lot smarter than I am.

On the other hand, starting out with a clean science slate did have its advantages. For one thing, even at a relatively advanced age there's a certain child-like joy in learning something new, especially when that something new is—to use the scientific term—really cool. Also, not having any preconceived notions spares you from stubbornly holding on to them. As this book will show, when it comes to Earth-impacting asteroids and comets, stubbornly held notions set back the cause of scientific progress for generations.

I should note that while this book is about a complex scientific issue, it's not really a science book. Instead, it's a story—a very human story about our long struggle to understand Earth's place in the Solar System, and the pivotal role of Earth-impacting asteroids and comets in shaping our world. Sure, there are scientists in this tale, men and women whose personal lives and characters are often just as fascinating as their scientific discoveries. But there are also cowboys and Indians and astronauts, Stone Age toolmakers and aerospace engineers, bold explorers and backyard amateur astronomers. King Tut and the mad teenaged Roman emperor Elagabalus make appearances in these pages; so do Teddy Roosevelt and the lead guitarist for the rock band Queen. On the technology side, picks and shovels and horse-driven machinery share the stage with space-deployed gravity tractors and laser ablation devices and ion beam shepherds.

In case you're wondering, yes, there are UFOs in this book—millions of them in fact. But these unidentified flying objects are asteroids and comets, not spaceships from faraway galaxies. I'll admit that given the breadth and scope of the universe, it seems likely that there is life somewhere beyond our own little planet. But with all due respect to the UFO belief community, when it comes to Earth visitations by vaguely humanoid space aliens, the science doesn't seem to be in on that one quite yet. Besides, as far as I'm concerned, the proven science behind Earth-approaching comets and asteroids is already out-of-this-world astonishing; we don't need to dress it up to make it more so. Same thing goes for astrology. I heartily agree that fate is written in the stars and planets. But the basic premise of this book is that the fate of Earth and our Solar System is determined by the collision of celestial bodies, not their alignment.

Finally, I should note that science, like the Earth itself, is a constantly moving target; sometimes it's hard to get a bead on it. Theories that were once scientific heresy have become scientific dogma, and vice versa—with the asteroid-related extinction of the dinosaurs being a prime example. Over the past almost four decades, hundreds of scientists have written literally thousands of scientific papers on that subject; entire forests have perished. And although the asteroid- or comet-impact extinction theory has largely carried the day, some scientists are still arguing about the issue, often in the most bitter personal terms. In this book I will chronicle that and other scientific controversies concerning Earth-impacting asteroids and comets, but I don't intend to litigate them. When there are competing theories on any given subject, I'll say so—and then I'll go with the one that currently seems to make the most sense.

With all that said, our story is waiting. It begins in the not-too-distant past, when another, much bigger asteroid came blazing through the Arizona skies. . . .

CHAPTER 1

IMPACT!

It's fifty thousand years ago, in the middle of a vast rolling plain near the future site of Flagstaff, Arizona. And high up in the sky, just on the edge of space, an asteroid half the size of a city block is hurtling toward this exact spot at 40,000 miles an hour.

For hundreds of millions of years this cold, jagged, half-million-ton piece of tumbling space metal has been orbiting around the Sun, staying in its own lane and not bothering anybody. But then some outside force—a collision with another asteroid, or some subtle gravitational nudge from another planet—sent it off on a new path, a new orbit. That orbit over time has conspired to place this speeding asteroid and our own speeding Earth in the exact same spot in space at precisely the same moment. And soon, very soon, this asteroid's traveling days will be over.

Earth isn't completely defenseless against this sort of attack from space. If this asteroid is actually going to hit the Earth's surface intact, first it has a gauntlet to run. The asteroid has to make it through the atmosphere.

You wouldn't think the atmosphere would pose much of a problem for a flying space boulder. After all, it's just a layer of air, and a pretty thin layer to boot. Proportionally, Earth's atmosphere is thinner than the skin of an apple. But air is like water. If you do a graceful swan dive off the low board at the pool the water provides a gentle cushion; but if you do a hundred-mile-an-hour belly-flop off the Golden Gate Bridge, it's almost like hitting concrete.

Same thing with this asteroid. It's a blunt object, with all the aerodynamic glide qualities of a blacksmith's anvil, and it's traveling twenty times faster than a rifle bullet. So when the asteroid hits the increasingly dense air of the atmosphere it gets some push-back. The air molecules in front of the asteroid can't get out of its way fast enough, and so the air in its path is radically compressed and heated up. Within seconds, temperatures on the asteroid's previously frigid surface reach three thousand degrees and more, lighting up both the asteroid and the wall of shock-compressed air in front of it in a brilliant incandescent glow, and leaving a fiery tail of bright light miles long. At the same time, the tremendous pressure causes pieces of the asteroid to break off and fall toward the Earth's surface on their own.

If our asteroid was smaller, or if it was made of rock instead of metal, the heat and pressure would probably destroy it, making it burn up or blow up before it ever reached the ground. Every day Earth's atmosphere easily destroys millions of incoming pieces of space debris, creating meteors that flash in the night sky. Most of those pieces of space debris are tiny, the size of a grain of sand, but even the bigger pieces usually fail to survive the atmospheric gauntlet. Our atmosphere wraps around the Earth like a thin sheet of Kevlar, protecting us from assaults by minor intruders from space—which is another reason we should probably take better care of it.

But this asteroid of fifty thousand years ago is no grain of sand, no minor intruder. This asteroid is 150 feet wide, and it's made of sterner stuff than rock. It's composed of almost pure nickel-iron, an incredibly heavy and strong alloy. A single cubic foot of it—about the size of a 50-pound block of ice—weighs almost as much as a

Harley-Davidson Sportster, and there are a million cubic feet of nickel-iron in this asteroid. Earth's atmosphere does its best, but completely destroying this asteroid is an impossible order. Nothing is going to stop this thing now. In just a few seconds it's going to slam into the Earth's surface like a giant cosmic cannonball.

But before that happens, let's freeze the asteroid in mid-plummet—say, about twenty miles above the Earth's surface—and take a look at what it's about to hit.

Ground zero for the incoming asteroid is a patch of mile-high tableland in what is now known to geologists as the Colorado Plateau. The climate is wetter and cooler on this day than it will be in modern times, the land a little more lush. Juniper and piñon pine woodlands are interspersed with grass-covered savannahs, and the ground is riven with flash-flood gullies and small flowing streams.

There are no people here on this day. The best scientific evidence suggests that humans won't arrive in North America for another thirty thousand years or so. Still, there is abundant animal life—and with one notable exception, much of the animal life would be familiar to us in modern times. There are beetles and sawflies and spiders, rattlesnakes and gopher snakes, pack rats, moles, and voles; in the air there are hawks and hummingbirds and swallows. The one notable exception is that on this day fifty thousand years ago the land is also populated with those large—in some cases magnificently large—and now extinct Late Pleistocene animals collectively known as megafauna.

One of them is a species of elephant-like mammoth known as *Mammuthus columbi*—named for no particularly good reason after Christopher Columbus—a 13-foot-tall, 20,000-pound giant that grazes in matriarchal family groups in the grassy open meadowlands. Another is a species of mastodon, a slightly smaller distant cousin of the mammoth that browses on coniferous twigs in the forests. There are ten-foot-long giant ground sloths in the area, as well as an ancient type of camel appropriately known as *Camelops hesternus*—Latin for "Yesterday's camel." There are herds of horses and bison grazing and galloping on the plains, but the horses are smaller than modern-day horses, and the bison are bigger than the American buffalo of later eras—much bigger, with enormous horns that measure eight feet from tip to tip. There are predators lurking about as well—packs of fearsome dire wolves, and saber-toothed big cats known as *Smilodons*. In the air, soaring on the thermals, there are giant condor-like birds with wingspans of 16 feet.

Mammoths and mastodons, pack rats and rattlesnakes, sloths and swallows and *Smilodons*—these are what are waiting for the asteroid now poised above their heads. Mercifully, the creatures in the impact zone have no idea what's coming. They can't hear the sonic boom the asteroid creates as it bursts through the atmosphere, because the asteroid travels faster than the sound it makes. And if they happen to look up at the northwestern sky, all they see is a bright glow, with no indication that this blazing rock is coming directly at them. So there is no sense of impending doom, no panic, no stampeding terror.

Well, it would be nice if this peaceful, bucolic Late Pleistocene scene could continue. But since it can't, we might as well get it over with. So now we go back up and unfreeze our plummeting asteroid—and two seconds later it hits the Earth with the force of a thousand Hiroshimas.

The impact isn't an explosion in the conventional sense. Instead, the destructive force comes from kinetic energy. Every moving object—a car on a freeway, a cue ball on a pool table, an asteroid hurtling through space—possesses kinetic energy, and the heavier the object is and the faster it's going, the more kinetic energy it has. As you can imagine, an asteroid that weighs hundreds of thousands of tons and is traveling at ten miles per second possesses an enormous amount of kinetic energy—and when it hits something it releases that energy with explosive force. To quantify that force, as a kind of shorthand scientists compare it to the energy released by a ton of TNT. The Hiroshima atomic bomb had the energy equivalent of about twelve thousand tons of exploding TNT. When our asteroid hits the ground, it releases the energy equivalent of some twelve *million* tons of TNT, or twelve megatons.

So even though the asteroid isn't a bomb, it might as well be. The impact shatters rocks a thousand feet below the surface and pulverizes millions of tons of stone into a fine, talc-like powder. The explosion peels up thick layers of ancient limestone and sandstone and folds them back on themselves, like a firecracker going off inside a stack of pancakes. It ejects almost two hundred million tons of rock into the air, hurling boulders the size of small houses hundreds of yards, and sending smaller pieces arcing upward in thousands of smoking contrails. A giant fireball starts rising into the sky, searing everything around it and creating a dusty brown mushroom cloud that reaches up to the stratosphere.

Obviously, any living thing in the asteroid's immediate point of impact—every tree, every bug, every *Smilodon* and *Camelops*—is instantly reduced to the molecular level. And those creatures are the lucky ones; they never know what hit them. Other animals farther from the impact point are less fortunate.

The asteroid's impact sends a shock wave radiating out in every direction, a burst of overpressure that causes internal organs to collapse and eardrums to burst and blood to bubble in veins. That's accompanied by a blast of wind moving in excess of 1,200 miles per hour—an atmospheric disturbance for which the word "wind" hardly seems adequate. This super-wind uproots trees, scours the ground bare of vegetation and sends everything in its path hurtling through the air in a maelstrom of debris—boulders, pebbles, jagged shafts of broken tree trunks, dead or dying rattlesnakes, pack rats, giant sloths. Twenty-thousand-pound mammoths are sent skittering along the ground like tumbleweeds.

Within three miles of the impact site, no living creature on the ground survives. Farther out the effects start to diminish, but even ten miles away the blast still hits with hurricane force, pelting victims with rocks and small debris and shredding their hides like a sandblaster. As if that weren't bad enough, for miles around the ground is bombarded by falling chunks of the asteroid that sloughed off in the atmosphere, and by millions of pieces of molten nickel-iron and other materials that were thrown up by the impact and are now falling back to the ground. It's a lethal rain of rock and iron.

And then, within just a few minutes, it's over, the only sound the whimpers and bellows of the wounded. What's left is a 300-square-mile patch of ground that has been scorched and stripped bare. And at the center there's a blackened, smoldering, bowl-shaped hole in the earth, a crater almost a mile across and 700 feet deep, surrounded by a rim of displaced rock some twelve stories high.

It's an astonishing amount of destruction. And yet, there's something about this asteroid you should know. You should know that in the cosmic scheme of things, our asteroid's violent collision with the Earth on this day fifty thousand years ago really isn't any big deal. It's not even all that unusual. The fact is that over the course of four and a half billion years, Earth has been hit *millions* of times by asteroids and comets as big as or bigger than this one—in some cases many orders of magnitude bigger. Earth has been bombarded by hurtling space bodies the size of mountains and even small planets, with collisions so powerful they tilted the Earth on its axis, sent huge chunks of Earth's surface flying into space and enveloped the globe in shrouds of fire and dust that wiped out most of its species.

Our relatively small asteroid has done none of those things. As tough as it's been for a few mastodons and giant sloths in the impact zone, other creatures grazing thirty miles away survive the asteroid's impact quite nicely; it doesn't even ruin their day. Within a few years, the blackened ground around the asteroid crater will be covered with new vegetation, and the animals will return. Except for the huge hole left in the ground, it will be almost as if it never even happened.

The bottom line is that by Earth-impacting-asteroid standards, our little asteroid of fifty thousand years ago has actually been something of a runt, a piddler, a weak sister.

So why do we care about this particular asteroid? What makes it so special?

The answer is simple. It's because someday—in about forty-nine thousand and nine hundred years, to be more precise—this little asteroid and the crater it has left behind will become the most important of their kind in the history of science.

They're going to change everything.

○

The crater is still there today. Officially it's called Meteor Crater—which, as we'll see, is a gross misnomer on several levels—and finding it is easy. Just head east or west on Interstate 40.

Westbound on I-40 from the New Mexico–Arizona border, the freeway traces the path of the old and fabled Route 66, the famous "Mother Road" that was an important part of American culture until it was strangled and mostly buried by the Interstate Highway System. The route passes through some of the most rugged and most beautiful country in the American Southwest. There are the magnificent badlands of the Painted Desert, so named by conquistadors under Francisco Coronado for the soaring buttes and mesas layered up in brilliant bands of lavender and vermillion and magenta; the vistas look like sunsets rendered in rock. The highway also transects the Petrified Forest National Park, a Mars-scape where fallen stands of giant pine trees more than two hundred million years old have been literally turned to stone—smoky quartz, purple amethyst, yellow citrine. The Petrified Forest is an eerie place, a haunt for ghosts.

There aren't any cities out this way, and the few towns along the route tend to be low-slung and dusty. Farther west on I-40 there's the small ranching and railroad community of Holbrook, which not so very long ago had its own brush with bombardment from space. In 1912 a space boulder about the size of an ice chest exploded high in the air east of town, pelting the ground below with thousands upon thousands of pea-sized meteorites—an event that was known as "the day it rained rocks." Or at least it used to be known that way. When I ask the waitress at a coffee shop about the meteorite fall, she's never heard of it. Oh? And has she lived here long? All my life, she says—which is another thing about assaults from space. We tend to forget them pretty quickly.

Still farther westward there's Winslow, a once-bustling Route 66 town now fallen on leaner times. If the name is familiar it's probably because of the line from the 1972 Eagles hit song, "Take It Easy": *Well I'm a standin' on a corner in Winslow, Arizona. . . .* That's just about the only thing Winslow is famous for. In fact, after the interstate bypassed the city center in the late 1970s, the Winslow town fathers tried to revive their downtown's flagging fortunes by building the Standin' on the Corner municipal park, complete with a flatbed Ford and a bronze statue of the late Eagles lead singer Glenn Frey. The effort has been a modest success. Every year tens of thousands of travelers, most of them graying Boomers, stop by to take selfies with the statue and browse in the nearby gift shop, where the sound system plays Eagles' songs—and nothing but Eagles' songs—all day long. "You get to where you don't even hear it anymore," a weary clerk tells me.

North of town I-40 continues west, straight as a rifle shot, through a vast expanse of high desert (elevation about 5,500 feet). This is not the woodland-savannah of fifty thousand years ago; sadly, there are no grazing mammoths or herds of galloping giant bison. A few stunted junipers struggle to survive in the washes, but other than that a guy could go blind looking for a tree. Broiling in the summer, frigid in winter, 20- and 30-mile-an-hour winds are a constant, and gusts of more than 100 miles per hour have been recorded. On the highway, 80,000-pound big rigs shimmy and sway in the crosswind; passing them can be an experience. Except for the highway and the Burlington Northern and Santa Fe train tracks and the occasional billboard touting tourist attractions that lie ahead—"Jewelry Made By Indians!" "Moccasins For the Whole Family!"—there is simply nothing here. It's the exact geographic center of the lonesome heart of nowhere. People who enjoy desolate places might find a stark minimalist beauty in it. Those who don't might just want to step on the gas.

Eastbound, the trip to Meteor Crater is a little more scenic, at least at first. First there's Flagstaff, an attractive mountain town of some seventy thousand souls that bills itself as the gateway to the Grand Canyon eighty miles north. Flagstaff is home to Northern Arizona University and the century-plus-old Lowell Observatory, a piney, campus-like facility that was the nerve center for countless space discoveries—including the discovery of thousands of fresh new asteroids, some of which could someday come our way. I'll get back to Lowell Observatory later.

East on I-40 there are mountains covered with stately Ponderosa pines, then rolling hills with copses of piñon pine. Twenty miles out from Flagstaff there's Winona, another old Route 66 small town best known for its brief mention in a popular song—in this case, the 1946 Nat King Cole hit, "(Get Your Kicks on) Route 66." *Flagstaff, Arizona / Don't forget Winona*, the song says, and when I repeat the lyric to the old-timer behind the counter at the gas station/convenience store he smiles and gives me a free "Don't Forget Winona" postcard. Farther along eastbound, back in the high desert now, on the right are the crumbled remains of a famous 1940s Route 66 tourist trap called the Twin Arrows Trading Post—the "arrows" being two telephone poles stuck at an angle in the ground with plywood sheets for fletchings. The arrows are still there, but the trading post name has since been appropriated by the nearby Twin Arrows Native American casino, where the Navajo Nation exacts its long overdue revenge. Although it's not visible from the highway, still farther on are the equally dissolute remains of an Old West ghost town called Canyon Diablo, which will play a role later in our story.

But whichever route taken, westbound or eastbound, eventually there is Exit 233. I steer my pickup onto the exit ramp, past the Meteor Mobil station and the Meteor RV Park, and then head south on a two-lane stretch of good blacktop. As signs along the road announce, I'm crossing part of the famous Bar T Bar Ranch, a family-owned spread of 300,000 acres of private and leased land that prides itself on raising range-fed, antibiotic- and growth-hormone-free Angus cattle. (If you'd like to eat one, you can do so at the upscale Diablo Burger restaurants in Flagstaff and Tucson.) The ground around here is still littered with small fragments of meteorite material left by our asteroid of fifty thousand years ago, but I can't pull over to look for any. As the same signs forcefully point out, it's private land, and the Bar T Bar hands keep their eyes peeled for trespassers.

I can't see the crater itself from the road, but I can see what's wrapped around it—a low, wide hill that rises above the surrounding plain. Over the past century and a half it's been variously described as a butte, a bluff, a mound, and a mountain—although it seems like a stretch to call a landform just 160 feet high a mountain. Whatever the appropriate term, perched near the top of its north slope is the Meteor Crater Visitor Center, a brick and glass and native sandstone structure designed in the 1960s by noted Modernist architect Philip C. Johnson. (The even more noted Frank Lloyd Wright also submitted a design, but he didn't get the gig.) Although Meteor Crater is an official U.S. Natural Landmark site, for reasons I'll get into later both the visitor center and the crater itself are privately owned. Admission (in 2018) is a hefty eighteen bucks, but there are discounts for youngsters and oldsters and military types; a flash of my veteran's ID card earns me a 50 percent discount and a cheerful "Thank you for your service!" from the cashier. The visitor center features a museum, interactive computer displays and an 80-seat theater/auditorium. There's also a Subway sandwich shop and a gift shop stocked with Meteor Crater T-shirts, Meteor Crater caps, Meteor Crater coffee mugs; for eight bucks I buy a tiny piece of the oxidized meteorite material that I'm not allowed to look for beside the road. For the most part, though, it's all very tasteful. No giant concrete dinosaurs, no garish signs—"See the Thing from Outer Space!" The atmosphere is sort of like a small-town community college campus.

But before I visit the museum or watch the short film on asteroids in the theater, first I want to see the crater itself. So I take the stairs up to the observation level on the crater rim, push my way through the glass doors and there it is, spread out below me.

The effect is—well, it's breathtaking. Literally. The first time I see it, it actually makes me gasp.

It's a struggle to convey the immensity of the thing. I could say that at 4,000 feet wide and 550 feet deep, with a flat bottom and steeply sloping walls, this huge hole in the earth is as deep as the Washington Monument is tall, that it would take fifty billion gallons of water to fill it up, that you could stack almost three million Greyhound buses inside it. Or try this. Imagine a football stadium that rather than holding the usual seventy or eighty thousand seats instead holds *two million* seats. Fans sitting in the nosebleed sections would be watching the game from half a mile away; they'd need a telescope to see the snap.

But no, none of that quite captures the effect. It really has to be experienced.

Some three hundred thousand people a year do experience it, taking the five-mile detour off I-40 in their RVs and SUVs and tour buses. Kids especially seem impressed; as they come boiling through the glass doors and see the huge crater for the first time, their "Awesomes!" and "Epics!" ring out over the wind, followed closely by their parents yelling at them not to climb on the guardrails. Many of the grownups seem stunned, as if they can't quite believe it. "This is the ninth wonder of the world," a guy carrying an enormous camera bag tells me—but then his wife calls him away before I can ask him what he thinks the eighth wonder is.

Inevitably, though, there's also some grousing. A woman in a red Nebraska Cornhuskers T-shirt declares—loudly—that "Eighteen dollars is a lot of money just to see a big hole in the ground!" It's a common complaint.

"Yeah, we hear that a lot," admits an affable young tour guide dressed in quasi-official khakis. "People get really mad when they find out their National Parks annual passes aren't any good here."

Most frequently asked question: Was this the one that killed the dinosaurs? Answer: Not even close. Even that seems to annoy some people, as if they've been lured into some kind of asteroidal bait-and-switch.

But the grumblers and grousers are sadly missing the point. True, Meteor Crater may not visually match the crimson-hued beauty of the Painted Desert, or the seemingly endless majesty of the Grand Canyon, or the spooky allure of the Petrified Forest. Meteor Crater *is* a big hole in the ground, rendered in earth tones of gray and tan and ochre. But what gives Meteor Crater its emotional power is not its inherent beauty. Instead, it's the manner of its creation.

The Grand Canyon and most other natural wonders of the American Southwest were formed over unimaginable depths of time, with wind and water scouring away the ground, grain by grain, over millions upon millions of years. They represent Nature at its most plodding and patient; they challenge our grasp of how long the Earth has been here, and how brief is our time upon the planet.

Meteor Crater, on the other hand, was created in just a few milliseconds of unexpected and almost inconceivable violence—and thus it represents Nature at its most capricious and destructive. It challenges us to think not just about what *did* happen, but what *could* happen, in an instant, literally out of the blue. Stand on the crater's rim, gaze across the vast expanse, imagine the flaming asteroid hurtling toward Earth, the ground-shattering explosion, the mushroom cloud rising 40,000 feet into the sky, the rain of iron and rock falling down for miles around. And then remember that the burst of violence that left this enormous scar on the Earth was relatively small in comparison with others. Multiply what happened here by a thousand times, and then maybe you'll begin to grasp the destructive potential of a major asteroid impact.

The crater hasn't changed much since the moment it was born. It's not quite as deep as it was back then—a few hundred feet of wind and water-borne sediments have collected at the bottom over the millennia—and the walls and rim are dotted with clumps of salt brush and riven with countless water-cut gullies; rain doesn't come often here, but when it does, it comes a gusher. But for the most part, the crater is still in geologically pristine condition—which is what makes it unique in all the world. Sure, there are other "impact craters" created by asteroids or comets that hit the Earth—a couple hundred of them that we know about, thousands upon thousands that we don't. But in most cases those other known impact craters have been worn away by erosion or covered by vegetation or filled in and scraped flat by glaciers. Their existence can be perceived by scientific instruments and scientific deduction, but most can't actually be seen.

But Meteor Crater can be seen, in almost the same state it was in fifty thousand years ago. There are a couple of reasons for that. One is the crater's relative youth. In geologic terms, fifty thousand years is a blink of an eye. There simply hasn't been time for this crater to be worn away or destroyed like so many others. The other reason it's in such good shape is its location in the arid high desert. It's almost as if the crater came with safe-handling instructions: *Store in a cool, dry place. Best if studied within fifty thousand years.*

And Meteor Crater has been studied, by generations of scientists who have trooped over its shaley slopes and chipped away at its rocks and peered through microscopes at the pieces of the asteroid that were scattered about. And slowly, over decades, the asteroid and the crater it left behind have given up their secrets. How old is the Earth? Why is the face of the Moon pockmarked with scars? What happened to the dinosaurs? Those are just a few of the questions that this crater and the asteroid that made it have helped us to answer. This crater has become a kind of astrogeological Rosetta Stone, a guide into the mysteries of our planet and our Solar System.

Perhaps even more important, this place has served as a kind of early-warning system, a canary in our coal mine, a Cassandra foretelling dire things to come. Until not so very long ago, we thought that our Solar System was a basically benign and predictable place; hardly any reputable scientists believed that large asteroids and comets had ever struck the Earth, certainly not with enough force and power to alter life on our planet. But now we know better. Because of the scientific revolution that this crater and that asteroid of fifty thousand years ago set in motion, we now know that Earth exists in a kind of cosmic pinball machine, a place of chaos and catastrophe and worlds in collision.

And as we'll see, we're finally taking that new knowledge seriously—although it's an open question whether we're taking it seriously enough. But at least we're trying. On windswept mountain peaks scattered around the world, astronomers are scanning the nighttime skies for asteroids that could pose a threat to Earth—and they're finding thousands of them that we never knew existed. NASA, the Federal Emergency Management Agency, and other national and international agencies are routinely holding "Planetary Defense Conferences" to try to assess the asteroid threat, and to come up with plans for what to do about it. Millions of miles away from Earth, space vehicles are conducting recon missions against potential enemy asteroids that have a small but still significant chance of colliding with Earth when our children's great-great-grandchildren are living their lives. Visionaries are imaging how we might turn those dangerous asteroids to our own advantage, in the process saving humanity from destruction by our own hands.

The point is that we owe a lot to this crater and our little asteroid of fifty thousand years ago. They have told us many things.

But what's astonishing is how long it took us to listen.

CHAPTER 2

MIRACULOUS APPARITIONS IN THE AYRE

In 1628 the English playwright and essayist Thomas Dekker published a short pamphlet titled "Looke Up and See Wonders: A Miraculous Apparition in the Ayre." In it Dekker described a recent mysterious disturbance in the sky above the small Berkshire village of Hatford.

"In an instant was heard first a hideous rumbling in the ayre," Dekker wrote, "and presently after followed a strange and fearful peal of thunder, running up and down these parts of the country. . . . All men were so terrified, that they fell on their knees, and not only thought, but sayd, that verily the Day of Judgment was come."

Based on eyewitness reports, Dekker described the event as a kind of artillery battle in the sky, complete with sounds like cannons booming and clouds of smoke and flying rocks "not unlike the flying of bullets from the mouths of great ordnance." Several of those rocks—Dekker called them "thunder-stones"—were later found on the ground, and were described by Dekker as "In color outwardly blackish, somewhat like iron. . . . Within it [there is] soft material, mixed with some kind of mineral, shining like small pieces of glass."

Dekker didn't know it, of course, but the rocks were meteorites, and the smoke and sounds of cannons were caused by a small asteroid that had exploded in the atmosphere. And his report about what happened in Hatford actually wasn't all that unusual. It was just one of many, many accounts of "miraculous apparitions in the ayre" that have been passed down through history.

Fireballs and meteors streaking through the atmosphere, stones falling out of thin air, comets setting the heavens ablaze—they have been part of the human experience since the first Homo sapiens gazed up at the African sky. And ever since, they have inspired fear, awe, reverence, even madness. Before we get on with our story, we should know what these miraculous apparitions are—and what they were long thought to be.

Because they're visible over vast areas of Earth, and because they can linger in the sky for weeks and even months, and because they're beautiful, comets have always attracted the most attention—and not always welcome attention.

Comets are loosely bound chunks of ice, dust and rock that were left over from the creation of the Solar System more than four and a half billion years ago. Some of those chunks are a few hundred yards wide, others tens of miles across. There are probably trillions of comets out there. Some of them reside in what's known as the Kuiper Belt, a doughnut-shaped region of space way out beyond the orbit of Neptune. Even greater swarms of comets are believed to make their homes in the Oort Cloud, a spherical assemblage of frozen space bodies that envelopes the farthest reaches of the Solar System.

How far away are these comet-rich regions? They're far, really far, up to trillions of miles from the Sun. To put those Solar System distances into perspective, try this mixed-sports demonstration. Take a regulation NBA basketball, paint it yellow and red and orange, then put it on the northernmost goal line of the MetLife football stadium in East Rutherford, New Jersey. The basketball is the Sun. Now take a single pellet of No. 8 skeet shot—less than a tenth of an inch in diameter—paint it blue and put it on the 28-yard line. That's Earth, and it shows you just how far our Earth is from the Sun, and how small Earth is in comparison; it's a tiny BB to a basketball. Meanwhile, Mars is a pinhead on the 47-yard line, and Jupiter, the biggest planet, is a Whopper malted milk ball high up in the stands behind the opposing goal post, 150 yards away from our basketball/Sun. The farthest edges of the Kuiper Belt are miles away, across the Hackensack River in Secaucus. And a comet at the outer edge of the Oort Cloud? It's a microscopic dust mote, and it's not even in New Jersey. It's more than 3,000 miles away, flying over Brazil.

Like I said, it's far.

When they're simply cruising around in the frigid vastness of the Kuiper Belt or the Oort Cloud, comets really aren't much to look at. They've been described as "icy snowballs" or "dirty iceballs," although in most cases they aren't actually balls. They're more like dirty snow clumps or icy dirt clods. Most comets stay in their own lanes out in deep space, going around the Sun in long, slow orbits, but occasionally some gravitational disturbance will put a comet into a highly elliptical (oval-shaped) orbit that sends it speeding into the inner Solar System. And that's when a comet literally starts to shine.

As a comet gets to within a few hundred million miles of the Sun it begins to warm up, and the frozen substances on and just below the surface—water ice, frozen methane, and so on—turn into gases mixed with dust particles. Those glowing, sunlight-reflecting clouds of dust and gases spew out of the comet like smoke from a 1969 Mercury Comet with a cracked block, forming a halo or "coma" around the comet's icy-dirty nucleus; sometimes they also form long, bright tails of dust and gases. A big gassy comet can have a coma a million miles across, and tails that stretch for hundreds of millions of miles through space.

Comets zip through the inner Solar System all the time, but most of them are too dim and too far away to be noticed by anyone except astronomers armed with telescopes. But every now and then there are comets that are big enough and bright enough and come close enough to Earth to light up vast swaths of our sky. Those are called Great Comets, and they are some of the most spectacular sights the heavens have to offer.

And yet, throughout most of human history Great Comets were not associated with beauty, but instead with impending doom. In cultures from ancient China to the Americas to the Middle East to Europe, they were almost universally denounced as "vile stars," harbingers of catastrophe, warnings from the heavens.

A German jingle written about the Great Comet of 1618 summed up the feeling this way:

> Eight things there be a Comet brings
> When it on high doth horrid range:
> Wind, Famine, Plague, and Death to Kings,
> War, Earthquakes, Floods, and Direful Change.

The alleged direful effects of comets were a staple for astrologers of the day. (Until the eighteenth century there was little distinction between astrology and astronomy in Western culture; some of the greatest astronomers of the time made their livings casting horoscopes.) For example, after pondering the Great Comet of 1664, one English astrologer offered up a menu of potential comet-induced disasters that included "fornication to become rife and common"—actually a pretty safe prediction in any era, comet or no comet—as well as "rotting of fruit," "abundance of caterpillars," "the death of some king," and the "destruction of fish." So fraught with evil were comets thought to be that one clergyman suggested they were physically composed of "the thick smoke of human sin . . . full of stench and horror"—which prompted another scholar to wryly note that if comets really were made of human sin, there should be a lot more comets.

Before we laugh at such cometary superstitions we should remember that even in relatively modern times comets have inspired madness and mayhem. For example, when a Great Comet appeared in 1910 a French scientist warned that when Earth passed through the comet's tail it would be enveloped in deadly cyanogen gas, which could "possibly snuff out all life on the planet." Other scientists shouted him down, but the report prompted panic in some cities, as well as brisk sales of gas masks and bogus "comet pills." More recently, in 1997, when the magnificent Comet Hale-Bopp was lighting up the sky, rumors spread that the comet was running interference for an alien spaceship that was following close behind. Believing this, thirty-nine members of the Heaven's Gate religious cult committed suicide so their souls could be transported to the UFO. The science of comets has come a long way, but apparently it's hard for human nature to catch up.

Anyway, it wasn't until 1705 that English mathematician and astronomer Edmond Halley managed to get a handle on the basic essence of comets. Based on German astronomer Johannes Kepler's laws of planetary motion and Isaac Newton's recently published laws of gravity, Halley was able to compute the specific orbit of the Great Comet of 1682. Although it can change over time, the orbit of a space object is sort of like a fingerprint: each is unique. And when Halley took a look at some old records, he found that the orbits of the Great Comets that had appeared in 1531 and 1607 bore an eerie resemblance to the orbit of the Great Comet of 1682. In other words, *they were all the same comet*, coming and going and coming 'round again in long orbits around the Sun.

Halley predicted that the Great Comet of 1682 would reappear after being gone for 76 years—which it did, on Christmas night in 1758, although Halley didn't live to see it. The reappearances of Halley's Comet in 1835 and 1910 were both spectacular. (The comet that sparked sales of "comet pills"? That was Halley's.) Unfortunately the 1986 Halley's Comet appearance was something of a visual bust, at least for casual observers on Earth; the comet and the Earth just weren't lined up right. Still, a series of space probes sent up by the European Space Agency, Russia, and Japan managed to gather a host of data on the comet's characteristics. Because of them, we now know that the ice-dust-rock nucleus of Halley's Comet is about eight miles long and that it's ingloriously shaped like a peanut. It's due back in 2061.

Halley didn't actually discover the comet that bears his name—it had been seen by people for thousands of years—but these days comets are named after the person or persons who spot them first. And you don't have to be a professional astronomer to claim naming rights on a comet. For example, Comet Hale-Bopp was separately but simultaneously discovered in 1995 by Alan Hale, a professional astronomer in New Mexico, and by Thomas Bopp, a parts department manager for an Arizona building supply company who was looking at the sky through a borrowed homemade telescope. In three hundred years of sky watching by thousands upon thousands of astronomers only about five thousand comets have been discovered, so discovering a comet is still a pretty big deal.

Seventy-six years may seem like a long time for a comet like Halley's to make one loop around the Sun—the Earth does it in just 365 days—but in the comet world that's actually a pretty quick trip. Some comets can take hundreds of thousands or even millions of years to complete their elliptical orbits around the Sun. There may be Earth-passing comets out there that were gawked at by *Australopithecus* man two million years ago, comets that won't be gawked at by another hominid until another million years from now—assuming there are any gawking hominids still left on Earth by then.

Not all comets survive their close encounters with the inner Solar System. Some of them crash into the Sun or other planets, or get slung out of the Solar System altogether, never to be seen again, at least not by us. Others just sort of wither away. Every time a comet passes close to the Sun it loses millions of tons of the dust and gases that form its coma and tail. Eventually, after hundreds or thousands of passes, the dust and gases are all played out, or they're trapped inside the comet's sooty, crusty surface. At that point the show is over, and the comet becomes just a dark dirt clod flying around in space. Even Halley's Comet is expected to fizzle out in twenty or thirty thousand years.

But there are plenty of others standing by to replace the fizzlers and light up the sky. As the astronomer Kepler once said, "There are as many comets as fishes in the sea." And the way we're going here on Earth, it seems likely we'll run out of fishes long before we run out of comets.

○

Meteors—also called shooting stars or falling stars—may not be as massively impressive as comets, but they can still put on a show. In fact, you can probably see a meteor or two this very night if you want to. Just find a spot where the sky is clear, preferably away from city lights and on a moonless night, then stretch out on a reclining lawn chair and look up. You have to be quick, though. Most meteors last only a second or two. They are literally gone in a flash.

A meteor is defined as the bright streak of light that's created when a piece of space rock—called a meteoroid—speeds through the atmosphere. Most meteoroids that hit the atmosphere are small, about the size of a pebble or a grain of sand, and yet they can make a flash that's miles long and visible from a hundred miles away. That's because they're incredibly speedy. Depending on the angle of impact, meteoroids hit the atmosphere at velocities of anywhere from seven to forty-five miles *every second*. Speed makes heat, which makes light, which makes a meteor.

Larger meteoroids that hit the atmosphere—pieces about the size of a baseball or bigger—create fireballs that can be seen over wider areas. Fireballs are basically just really bright meteors. As a fireball-sized meteoroid burns its way through the atmosphere, the heat and air pressure can sometimes cause it to brilliantly blow up into a million little pieces, at which point it's called a bolide. The fireball that I saw explode over Arizona in 2016? That was a bolide.

Like comets, meteors were long thought to be atmospheric phenomena. Aristotle linked them with clouds, lightning, thunder and other things "high in the air," the ancient Greek word for which was *meteoros*. As a result, the study of things high in the air eventually became the science of meteorology—which is why your local TV Action News weather guy is called a "meteorologist," while people who actually study meteors and meteorites have to make do with the ungainly "meteoriticists."

Also like comets, there's no shortage of meteors and fireballs. The fact is that millions of meteors blaze through Earth's atmosphere every single day, although most of them are never seen by human eyes. They're either obscured by clouds, rendered invisible by bright sunlight or flash over unpopulated areas, such as the 70 percent of the globe that's covered by ocean. And sometimes they're just gone too quickly for anybody to notice.

Almost all of the pieces of space debris that create meteors burn up completely before they come anywhere near Earth's surface. But a small percentage of space rocks do manage to make it through the atmosphere and hit the ground at least partially intact—at which point they're called "meteorites."

Some meteorites are made mostly of rock, some a mixture of rock and metal, and a few are almost pure metal, primarily nickel-iron. Most meteorites are small, less than an ounce, but there have been some monsters that survived their fall to Earth. The biggest yet found is the Hoba meteorite, a 64-ton chunk of space metal that was discovered in 1920 in South West Africa (now Namibia) by a farmer who was plowing a field and hit something that went *clank*!

How many meteorites hit the Earth is hard to say. One estimate suggests that up to eighty thousand meteorites bigger than a piece of gravel hit Earth's surface every year, although only a tiny percentage of them are ever found. Which makes you wonder: With all these meteorites falling from the sky every year, why don't more of us get conked on the head by them? Mostly it's because Earth is a big place, with about 200 million square miles of surface area. If you do the math, it works out to an average of one plummeting meteorite for every 2,500 square miles per year. Unless you're feeling really unlucky, it's probably safe to step out of the house without putting on a hard hat.

Still, people do get hit by falling space rocks, or at least tiny pieces of them. Every year tens of thousands of tons worth of microscopic "cosmic dust" from meteors, comets and outer space objects waft down through the atmosphere and land on us. Run a finger over the top of your refrigerator and what you'll find are a few microscopic bits of space dust mixed in with the ordinary Earth dust.

As for people getting hit by bigger chunks, that happens too, although it's rare. One of the most famous such incidents in recent history happened when a small fireball exploded over the town of Sylacauga, Alabama, in 1954. An eight-pound piece of the bolide crashed through the roof of a small rental house, bounced off a radio console and hit a 34-year-old woman named Ann Hodges on the left hip as she was sleeping on the couch, leaving a nasty welt. The event was a sensation, with Mrs. Hodges being hailed as "the first person in history to be hit by a meteorite." She was featured in *Life* magazine and appeared on the TV quiz show *I've Got a Secret*, for which she received $80 and a carton of Cavalier cigarettes. But sadly, after a drawn-out legal battle over ownership of the meteorite and a subsequent nervous breakdown, this shy and unsophisticated woman died at the relatively young age of fifty-two. If you'd like to see the meteorite that hit her, it's on display at the Alabama Museum of Natural History in Tuscaloosa.

(By the way, if a meteorite crashes through your roof you don't need a special meteorite rider on your homeowner's insurance policy to cover it. According to the Insurance Information Institute, your standard policy would cover meteorite impacts under a general "falling objects" clause.)

Fatal impacts of meteorites on human heads are even more rare—so much so that some science writers glibly claim that in all of recorded history there has never been a confirmed case of a person being killed by a meteorite. The key word here is "confirmed." Researchers who have actually studied the records have come up with a plausible list of scores of people killed by meteorites. This includes a dozen people killed by a meteorite in seventh-century China, thousands possibly injured or killed by a giant explosion and meteorite shower in China in 1490, a Milanese monk mortally wounded by a meteorite in 1664, a French farmer crushed to death by a meteorite in 1790—and on and on. True, a monk here and a farmer there doesn't exactly constitute a hellish rain of death from the sky. But despite what you may have heard, meteorites have killed people—and they inevitably will kill again.

Unfortunately, Western science was slow off the mark when it came to meteorites. Despite centuries of reports, such as the one Thomas Dekker wrote about events in Hatford in 1628, most scientists of the day didn't believe that such things even existed. It wasn't until the 1790s that a rather homely, sad-looking German lawyer-turned-scientist named Ernst Chladni took a serious look at rocks falling from the sky. After extensive study of fireball sightings and the nature of some unusual rocks, including a 1,500-pounder found in Russia, Chladni published a report concluding that the rocks clearly had extraterrestrial origins—a report that later earned him the title of "Father of Meteoritics." Chladni's report was ridiculed by some, but Nature quickly backed him up with fresh evidence—most importantly, an exploding fireball in 1803 that scattered thousands of meteorites over a small French village in Normandy. French scientists hastened to the scene and finally concluded that yes, these things were definitely un-Earthly.

But that still left the question: From where in space did meteorites come? One popular theory was that they were spit out by volcanos on the Moon and then sucked down to Earth by gravity. (Back then it was assumed that the thousands of visible craters on the Moon's face were all volcanic; more on that misassumption later.) Others thought they were rogue pieces of interstellar space rocks. At the time, no one understood that almost all meteorites are pieces of the least glamorous—and in some ways the most dangerous—denizens of our own Solar System.

Which is to say, asteroids.

○

Most of us grew up with asteroids. The tiny Asteroid B 612 in Antoine de Saint-Exupéry's *The Little Prince*. Han Solo boldly taking his spaceship into an "asteroid field" to escape pursuers in *The Empire Strikes Back*. The long succession of Hollywood movies—*Meteor* (1979), *Armageddon* (1998), *Deep Impact* (also

1998), among many others—that featured killer asteroids or comets hurtling toward Earth. And of course there were those many, many wasted hours—and countless wasted quarters—spent blasting space rocks on Atari's *Asteroids* video game in bars and arcades. For us, asteroids are a familiar part of our cultural universe.

But in fact it's only been in the relatively recent past that humans realized that asteroids even existed—much less that they occasionally smack into Earth.

And we stumbled onto asteroids only because we were looking for something more spectacular—that is, a missing planet.

In the late eighteenth century astronomers noticed that there seemed to be a certain mathematical symmetry in the placement of the six then-known planets—Mercury, Venus, Earth, Mars, Jupiter, and Saturn. The only flaw in that mathematical formula was the disconcertingly wide and seemingly empty gap between the orbits of Mars and Jupiter. There should have been a planet there.

So a group of European astronomers got together to launch a concerted search for the "missing planet," with each astronomer assigned to search a particular section of the night sky. The astronomers called themselves—I'm not kidding—the "Himmelspolizei," or the "Celestial Police." On the night of January 1, 1801, one of them, an Italian priest named Giuseppe Piazzi, peered through his telescope and spotted what seemed to be the missing planet in the Mars-Jupiter gap. He named the new planet Ceres, after the Roman goddess of agriculture.

Unfortunately for Piazzi, there were a couple of problems with his new planet. For one thing, it wasn't alone. A year later another Celestial cop stumbled onto yet another "planet" in the Mars-Jupiter gap, which he named Pallas. The other problem was that the new planets were tiny. It turned out that Ceres was only about 600 miles in diameter, and Pallas only a little over half that. (By comparison, the smallest actual then-known planet, Mercury, is about 3,000 miles in diameter.) The new space bodies were clearly too small to rate as planets, so English astronomer William Herschel came up with a new name for them. Because from telescopes on Earth they looked like dim stars, he used the word "asteroid," ancient Greek for "starlike." (Today asteroids are also known as "minor planets.")

By 1807 the Celestial Police had found two more asteroids, Juno and Vesta, in the same area as the first two. (A phrase keeps running through my mind: *Freeze asteroid! Celestial Police!*) They speculated that there might be more asteroids in the vicinity—perhaps as many as a dozen!—but with the telescopes of the day they could only spot the biggest ones. And they had no idea what they were made of, or where they came from.

Well, we now know that asteroids are pieces of rock or metal that, like comets, were left over after the formation of the Solar System about 4.6 billion years ago. After the planets formed, these leftovers, some quite large, never quite managed to coalesce into planets themselves. Instead they were trapped in the empty space between Mars and Jupiter that had so confounded the Celestial Police—and as they hurtled around in that vast space they inevitably collided with one another. It was like throwing a bunch of bricks into a high-speed revolving cement mixer; over millions and billions of years these large objects kept smashing into one another to form smaller pieces, which in turn smashed into one another to form even smaller pieces, and so on. Those pieces are the asteroids.

As for the Celestial cops' estimate that there might be as many as a dozen asteroids, that was a little off. There are actually millions upon millions of these things flying around in orbit around the Sun, a few of them hundreds of miles in diameter but most ranging from tens of miles to a single yard wide. (If they're smaller than that they're usually classified as meteoroids.)

Like I said, as space objects go, most asteroids aren't very glamorous. Unlike comets, asteroids don't release shining clouds of dust and gases to light up our skies as they pass by. (Although asteroids and comets generally differ in composition, some asteroids are believed to be old comets that have lost their fizz.) Only a few of the biggest or closest-passing asteroids can ever be seen with the naked eye, and then only very rarely. Depending on their composition, asteroids can range in color from light gray to almost black; some asteroids look like lumps of charcoal. Although a few astcroids are roughly spherical, most come in an endless variety of irregular shapes, resembling peanuts, human molars, dumbbells, anvils, ladies' high-heeled shoes, and, most commonly, potatoes; in fact, some of the model asteroids used in making *The Empire Strikes Back* actually were russet potatoes. In 2000 a massive radio telescope in Puerto Rico discovered that asteroid 216 Kleopatra, which is the size of New Jersey, is shaped exactly like a giant dog-bone chew-toy.

Some asteroids are solid, while others are "rubble piles" of small pieces loosely held together by gravity. Some are made of rock, some a loose mixture of carbon, rock, and minerals, and some are almost pure metal. There's an asteroid out there named 16 Psyche that's 120 miles wide and is composed of nearly pure nickel-iron; it's believed to be the solid metal core of a larger space body whose softer surface layers were blasted off by collisions. Psyche is like a huge hunk of stainless steel flying through space; if you could hit it with a billion-pound hammer, and if sound could travel through space, it would ring through the Solar System like a giant gong. NASA is planning to send up an $850 million unmanned spacecraft to take a gander at Psyche in the next few years.

A surprising number of asteroids have their own satellites—little moons—that orbit around them, while others are double asteroids that spin circles around each other as they revolve around the Sun, sort of like a celestial pas de deux. Like the planets, asteroids rotate as they hurtle through space, although they do it in weird ways. Some rotate slowly, like chickens on a rotisserie, while others tumble rapidly end-over-end like a football kicked for a field goal.

When seen through even the most powerful Earth-based telescopes, asteroids look like little blobs of light. It wasn't until 1991 that we humans got our first close-up look at one. In that year the unmanned NASA spacecraft *Galileo* flew by ten-mile-wide asteroid 951 Gaspra, and two years later it passed by 37-mile-wide asteroid 243 Ida. Photographs sent back to Earth revealed that both asteroids showed scars from collisions with other asteroids: deep craters, chasm-like gouges, giant cracks. Most other asteroids have similar impact scars—and sometimes they can produce odd visual effects. For example, a half-mile-wide asteroid called 2015 TB_{145} that passed by Earth in 2015 looks for all the world like a human skull, complete with a gouged-out gaping mouth, eye sockets, and nasal cavity. The fact that 2015 TB_{145} passed close to Earth on October 31—Halloween—made it creepier still.

Most asteroids reside in the so-called asteroid main belt between the orbits of Mars and Jupiter—the place where Piazzi first discovered Ceres—but the asteroid belt isn't even close to the deadly asteroid field from the *Star Wars* film. Space is a big place—that's why they call it "space"—so even though there are millions of asteroids in the main belt there's plenty of room for them to spread out. We've sent numerous space vehicles speeding through the asteroid belt without ever hitting one.

As long as they stay in their own lanes in the main belt, asteroids don't pose any threat to Earth. But there are some asteroids that have been bumped or gravitationally nudged out of the main belt and sent into different orbits around the Sun, orbits that come close to Earth's own path through space. It's like a NASCAR racer that veers off the track and starts cutting doughnuts in the infield. Those asteroids are called "Near-Earth Objects," or NEOs, and there are a lot of them—almost twenty thousand that we know about, thousands upon thousands of mostly small ones that we don't know about.

Most of the Near-Earth Objects that whiz past Earth miss us by millions of miles—although in space terms a million miles is just a whisker. But sometimes they come a lot closer. For example, in 1989 an astronomer spotted a quarter-mile-wide asteroid named 4581 Asclepius that passed within 425,000 miles of Earth—what the astronomer described as a "close call" in cosmic terms. What made it especially startling was that the asteroid was spotted only *after* it passed by Earth, not before; if it had been on course to hit us we wouldn't have seen it coming. (According to NationalDayCalendar.com, which lists commemorative days that are suggested by businesses and organizations, Asclepius's close passage on March 23, 1989, supposedly inspired the observance each March 23 of "National Near-Miss Day," a day dedicated to close calls, although no one seems to know who first suggested the idea. I should note that according to the website, March 23 is also "National Melba Toast Day" and "National Chip and Dip Day.") To cite a more recent close call, in 2013 a 100-foot-wide asteroid called 367943 Duende missed Earth by just 17,000 miles. That's a whisker of a whisker; that asteroid came closer to Earth than the satellite that sends signals to my DISH TV. Proportionally that's like having a bullet miss your head by sixteen inches. It doesn't kill you or even hurt you, but if you have any brains at all it certainly should get your attention.

With improved asteroid detection and tracking systems, in recent years the number of reported close passes by asteroids has grown dramatically. In the first five months of 2018 more than thirty relatively small asteroids were detected passing by Earth by margins less than the distance between the Earth and the Moon—which is to say, they missed Earth by less than 240,000 miles. And those were just the ones that were spotted. Many more zipped close by without anyone on Earth even knowing about them. The problem is that because asteroids don't reflect much sunlight, they're hard to spot even with powerful telescopes, especially the smaller ones—which may be a good thing for our peace of mind. If we could somehow mount a giant porch light on every Near-Earth asteroid that's zipping around out there, the nighttime sky would look like the Fourth of July. It could be a little scary.

Predicting close misses by asteroids can sometimes be a tricky business. For example, in 1998 the director of the International Astronomical Union's Minor Planet Center, the organization that keeps track of asteroids, caused a stir when he announced that a newly discovered half-mile-wide asteroid called 1997 XF_{11} would pass within 30,000 miles of Earth in the year 2028—and then he added that there was "a small [but] not entirely out of the question possibility" that it would actually hit our planet, with catastrophic results. Jet Propulsion Laboratory scientists Donald Yeomans and Paul Chodas quickly discounted this estimate—1997 XF_{11} will actually miss us by 600,000 miles in 2028—but not before the initial report prompted a spate of end-of-the-world news stories. The most notable was in the *New York Post*, which ran a front-page headline in Second Coming–size type: "KISS YOUR ASTEROID GOODBYE!" The *Post* also conducted a man-on-the-street survey, asking people what they would do if the asteroid were going to hit tomorrow. Two-thirds of them said they would get drunk.

Asteroid close approaches provide abundant fodder for apocalypse prognosticators. Take asteroid 4179 Toutatis. A three-mile-long piece of rock that's shaped like a mallet, Toutatis is the biggest known asteroid that crosses Earth's orbit. Toutatis made close approaches to Earth—within a million miles or more—in 2004, 2008, and 2012, and each time there were Internet and radio talk-show rumors that despite what the boys at NASA were telling us, it was going to hit Earth and destroy the planet. Many people believed this. Before Toutatis's 2012 approach, NASA reported getting up to 300 calls a week from worried people demanding to know if the world was going to end—and that was just the people with enough on the ball to find NASA's phone number. Toutatis's next significant close approach is scheduled for 2069, and you can bet that along about 2068 the rumors will begin anew.

Of course, asteroids don't always miss Earth. In a sense, they hit Earth every single day. Those tens of thousands of meteorites that rain down on Earth every year? Almost all of them are small pieces of asteroids that managed to find themselves in Earth's path. As for the bigger ones, a couple of times a month an asteroid the size of a MINI Cooper blows up in the atmosphere with a force equal to hundreds of tons of TNT. An asteroid or small comet capable of destroying a city slams into Earth's atmosphere at least a couple of times every century. As for the big boys, asteroids a mile or so in diameter, big enough to cover a continent with fire and ash, one of them will probably smack into us every half-million years or so. And as we'll see, every once in a long, long while, an asteroid or comet nucleus the size of a mountain changes Earth's history forever.

Obviously, those are just probabilities, not predictions. A big asteroid or comet could come barreling toward Earth at any time, regardless of how long it's been since the last one hit. The only thing that's certain is that it will happen someday. Think of it this way: Imagine two blind bumblebees randomly buzzing around in the Superdome in New Orleans—which, like the Solar System, is a vast but still finite space. You probably shouldn't bet that the two bees will bump into each other on the first day, or the second, or the third. But did I mention that these bees are immortal? Given that, you can bet the family homestead that eventually, maybe next week, maybe ten years from now, those two little blind buzzing bees will collide. It's been said that time transforms the improbable into the certain, that anything that *can* happen *will* happen, given enough time. That's the way it is with large Earth-impacting asteroids.

As for how much wallop a particular asteroid can deliver, size certainly matters. So does density, whether the asteroid is solid metal or a relatively loose conglomeration of rock. But speed is the main thing. As I mentioned in the introduction, an asteroid's kinetic energy is expressed as one-half the mass times the velocity *squared*, so the faster it's going, the more powerful the punch. Asteroids and comets and the Earth all orbit around the Sun at speeds of many tens of thousands of miles per hour, so if one hits us more or less head-on the combined impact speed—and thus the destructive power—would be much greater than if it caught up with us from behind and then hit us. It's the difference between getting slammed head-on by a wrong-way driver on the freeway and getting bumped by a tailgater when you're braking for a squirrel.

How many asteroids have been discovered? Oh, about three-quarters of a million of them so far. Of course, that's just a tiny fraction of the total, and it's mostly just the bigger ones, a mile or more in diameter. But it's still a lot, especially when you consider that almost all of those asteroid discoveries were made in just the past twenty years. In fact, there have been so many asteroid discoveries in recent years that we seem to have trouble finding suitable names for them.

You may have noticed that some of the specific asteroids mentioned so far have names like Asclepius and Duende and Psyche, while others have to get by with just letters and numbers like 1997 XF_{11}. That's because when a new asteroid is discovered and confirmed it's assigned a number/letter designation by the International Astronomical Union (IAU) that reveals the year and order of discovery. After that the discoverer is allowed to give the asteroid a name of his or her choosing—but with so many new asteroids being discovered, sometimes the discoverer just doesn't get around to it.

At first naming asteroids was easy enough. By common agreement among nineteenth-century astronomers—at least the ones in Europe and North America; nobody checked with Asia or Africa—asteroids were always given names from Greek or Roman mythology, along with a number indicating their order of discovery: 1 Ceres, 2 Pallas, 3 Juno, etc. Then in the twentieth century asteroid discoverers started naming them after astronomers and other scientists (2000 Herschel and 2001 Einstein and 7000 Marie Curie) and after famous writers (2985 Shakespeare and 2999 Dante and 3412 Kafka). In more recent years asteroid discoverers have begun naming asteroids after movie stars, fictional characters, rock bands, athletes—the sky's the limit. There's asteroid 12818 TomHanks, 9007 JamesBond, 13681 MontyPython, 26858 MisterRogers, 4319 JackieRobinson. There's asteroid 19367 PinkFloyd, 3834 Zappafrank, 19383 RollingStones. The Beatles have asteroid 8749 Beatles, as well as individual asteroids named Lennon, McCartney, Harrison, and Starr.

But you don't have to be famous—or even a real or fictional human—to have an asteroid named after you. There are asteroids named after wives, girlfriends, cousins, and all of the finalists in the 2005 Discovery Channel Young Scientists Challenge middle school science competition. There are asteroids named after countries, states, cities. There is an asteroid named 88705 Potato. Altogether some twenty thousand asteroids have been named, but that still leaves hundreds of thousands of them waiting for names.

There are some rules for naming asteroids. For example, Near-Earth asteroids are still usually named after mythological figures from various cultures—Apollo, Amor, Toutatis, and so on. The IAU also insists that asteroid names be non-offensive—no obscenities or insults in any language—and they can't be named for any military or political figure who hasn't been dead for at least a century. There are asteroids named after George Washington and Ben Franklin and Karl Marx, but there won't be an asteroid named Trump, at least not for a while. Naming asteroids after pets is officially discouraged, although it has been done. A cat named after a *Star Trek* character is now immortalized by asteroid 2309 Mr. Spock—this to the horror of some people who care about these things.

"You shouldn't name an asteroid after a cat," one astronomer told me with genuine outrage. "It trivializes the entire process."

(I should note that this asteroid-naming process has no connection with those private companies who'll let you name a star for $19.95 and send you an "official" certificate. That's all well and good; nobody owns the stars, or the asteroids either, and anybody can call them whatever they want. But the IAU is the only internationally recognized organization charged with officially naming objects in space. And you can't buy an asteroid name from them.)

Of course, all of this information about asteroids would have been news to the Celestial Police and the nineteenth-century astronomers who followed them. They had no idea how many asteroids were out there, or what they were made of, or that they zip past Earth all the time, at distances that can only be described as shockingly close.

As the nineteenth century ended, astronomers began using long-exposure photographic plates to detect the sunlight reflected by asteroids, which made spotting them quicker and easier than visual observations. By 1900 astronomers had discovered some four hundred asteroids, but they still didn't know much about the asteroids they were discovering—and frankly, most astronomers didn't care to know. There may have been a sense of scientific accomplishment in discovering the first asteroid, or maybe even the fiftieth. But really, how much glory was there in finding the three hundred and twenty-seventh asteroid? And if an astronomer did stumble onto yet another asteroid, he was more or less honor bound to go through the laborious process of calculating its orbit, which took him away from more important astronomical matters, like stars. Close, fast-moving asteroids left messy, star-obscuring streaks on their photographic plates and generally mucked things up. To put it bluntly, for your average late-nineteenth- and early-twentieth-century astronomer, asteroids were a pain in the ass. Things got to the point where one German astronomer famously referred to asteroids as "vermin of the sky."

As for the idea that asteroids sometimes collide with Earth, well, in the late-nineteenth-century view that was absurd. Sure, our planet might occasionally get hit by a meteorite a dozen or so feet wide. But a full-blown asteroid? No way. All of the then-known asteroids were safely tucked away in the asteroid main belt, hundreds of millions of miles away from Earth; the notion that one of them might hit Earth was about as likely as Earth accidentally bumping into Mars. In all of recorded human history no one had ever seen a large asteroid hit the Earth or anything else. And even if asteroids had hit Earth in the far distant and unrecorded past, where was the evidence?

Actually, there was evidence that Earth had been routinely bombarded by asteroids. The evidence came in the form of a mile-wide crater in the Arizona high desert, a crater just waiting for someone to recognize it for what it was.

And we can pinpoint with precision when that process of recognition began. It began on a warm evening in Tucson in October 1902, when a man named D. M. Barringer stepped out onto the veranda of the San Xavier Hotel to enjoy a fine cigar.

CHAPTER 3

ASTEROID MINERS

Daniel Moreau Barringer. A sepia-toned photographic portrait from the time shows an unsmiling man of middle age, wearing a three-piece tailored suit with a Princeton school tie and a gold watch fob looped over his ample belly. He's a portly gentleman, and by our health-conscious standards he might be considered obese, perhaps even morbidly so. In the early twentieth century, however, the barrel-shaped male torso was considered a mark of substance, strength, power. Think Teddy Roosevelt, or William Howard Taft.

But what you notice most about the man in the photograph is his eyes. Wide-set and dark, they bore into the camera, and us, with a fierce, almost frightening intensity. Even from across a century, it's as if he's daring us to disagree with him.

Born in 1860 into a prominent but not particularly wealthy North Carolina family—his father was a congressman and a U.S. minister to Spain—Barringer was orphaned at an early age and raised by a brother in Philadelphia. After graduating from Princeton in the Class of 1879, he practiced law for several years before seeking his fortune in the mining business—this despite the fact that when it came to geology and mining engineering, he was largely self-taught.

In 1896 Barringer stumbled onto a small but potentially lucrative mining operation in southeastern Arizona, and although he lacked significant money of his own, he managed to assemble a consortium of wealthy investors to buy out the mining claim and started bringing in the men and machinery needed for a large-scale industrial mining operation. The gamble paid off. By the time the richest ore played out, the fabulous Commonwealth Mine had coughed up $10 million in gold and silver (almost $300 million today), a significant portion of which wound up in Barringer's pockets.

So by age forty-two Barringer was a wealthy man—wealthy enough to build a white-columned, antebellum-South-style mansion on a 37-acre estate on Philadelphia's prestigious Main Line for his wife and their three (eventually eight) children. He also owned lucrative interests in other gold- and iron-mining operations in Mexico and the western United States, and he served on the boards of several large companies. He belonged to all the best "gentlemen's clubs"—the term meant something far different than it does today—and he was also an early member of the Boone and Crockett Club, an organization of wealthy and influential big-game hunters whose illustrious ranks included then-president Teddy Roosevelt.

As for Barringer's personality, the reviews are mixed. Certainly he was convivial, fond of good conversation, fine cigars and whiskey, and—obviously—good food. His wife and children adored him, and he had a wide circle of friends who admired and respected him. On the other hand, Barringer could be stubborn, arrogant, overbearing, suspicious, petulant, and he had a hair-trigger temper that he wielded like a gun. When writing to or about his enemies—which is to say, anyone who didn't see things his way—the yellowing pages of his voluminous correspondence drip with venom and bile. His opponents are "fools," "damn fools," "ignoramuses," "liars," "asses." Their ideas are "demented," "absurd," "childish," "insane," "contemptible," "preposterous," "asinine." And on and on. He was, in a word, difficult.

It was during a business trip to the Commonwealth Mine that this difficult man found himself smoking a cigar on the porch of the San Xavier Hotel in Tucson. As he was enjoying the evening air he struck up a conversation with a slim, somewhat mournful-looking individual named Samuel J. Holsinger, who had a tale to tell about a giant crater surrounded by meteorites from space.

Holsinger was a 43-year-old former lawyer and newspaper reporter who at the time was working as a "special agent" for the federal General Land Office, the forerunner of today's Bureau of Land Management. His job was to patrol government-owned lands, usually on horseback, searching for squatters, illegal timber cutters, gypsy miners; he was sort of like an early-twentieth-century EPA cop. Holsinger told Barringer that while investigating some malfeasance in northern Arizona he had heard about this place called Coon Mountain, which wasn't really a mountain but instead was a giant crater, a mile across and hundreds of feet deep. Holsinger had talked to an Indian trader at the little town of Canyon Diablo who told him that the ground around the crater was littered with pure nickel-iron meteorites, some of them quite large. And here was the kicker: The Indian trader and other locals believed that the massive crater had been blasted out of the ground by a huge meteor or asteroid that had struck the Earth sometime in the distant past.

Barringer was so astonished by this story that he dropped his smoldering cigar.

"He besieged me with questions," Holsinger later recalled. "I related all I knew concerning the meteoric field, and he urged me to make further inquiries."

Every life has its turning points, and that casual conversation on the hotel porch certainly was a major one for Daniel Barringer.

It was at that moment that he came up with an idea, an almost breathtakingly audacious idea, one that he would pursue for the next quarter century. He would pursue it to the point of obsession—and in the end it would ruin him.

I'll get back to Barringer and his big idea in a moment. But first I should note that it's a little odd that he was so surprised by this tale of a giant crater surrounded by meteorites. It wasn't as if the crater had been any big secret.

There are a lot of stories about who first "discovered" what came to be known as Meteor Crater. The tale the tour guides tell the tourists is that in 1873 a U.S. Army cavalry scout named Charles Albert Franklin—aka "Buckskin Charlie"—stumbled onto the enormous crater while guiding an Army surveying expedition and was awestruck by the enormity of what he had found. It's a good story, but unfortunately, for reasons too complicated to go into here, it's also almost certainly bogus. (See the chapter notes.)

But maybe it doesn't matter. Whoever may have been the first European-American to see the giant crater—perhaps a weary conquistador in Coronado's wandering, gold-mad army, or some lone prospector whose name is lost to history—he certainly couldn't have discovered it, since Native Americans in the region had long been aware of its existence. Arrow points, potsherds and even a couple of Native American pit houses have been found near the crater's rim, some of them dating back half a millennium or more, long before Buckskin Charlie or any other white man ever came along. Whether the crater itself was of any particular significance to ancient Native Americans is unknown, but what was significant was what they found around it—that is, the meteorites.

You'll recall that as our asteroid blazed through the atmosphere the tremendous heat and pressure caused pieces of it to flake off and fall to the ground as meteorites, while other pieces were ejected by the impact explosion—and all of those pieces were almost pure nickel-iron. Nickel-iron is an alloy that's extremely rare on Earth's surface, so when pre–Iron Age Native Americans found those heavy, extremely hard metal pieces on the ground, they knew they were something special, something worthy of respect. One 135-pound meteorite that came from the asteroid was later found a hundred miles away from the crater amid the ruins of a tenth-century Native American village, wrapped in a turkey feather blanket and interred in a ceremonial stone crypt. Another eight-pound meteorite from the asteroid was found in similar Native American ruins some eighty miles away. Other pieces of the asteroid have turned up as far away as Mexico, indicating that they were widely traded among ancient Native American groups.

White settlers who arrived in the vicinity of the Arizona crater in the late nineteenth century also came to venerate the meteorites found around the crater, although for a different reason—they were worth good money. Ever since the existence of rocks from space had first been established in the early nineteenth century, a small but avid group of scientists, museum curators and private collectors had been buying and selling meteorites, often at substantial prices. These particular meteorites scattered around the Arizona crater would prove to have a similar appeal.

In the mid-1880s some Mexican-American sheepherders were grazing their flocks near what was called Coon Mountain when they noticed some unusual rocks half-buried in the ground. (Raccoons were actually as rare as mastodons on Coon Mountain; the landform was named after a local rancher by the name of Coon.) Underneath their crusty surfaces the strange rocks had a certain metallic sheen, and at first the sheepherders thought they had stumbled onto chunks of silver. Later they showed the rocks to the sheep-camp cook, a former prospector who decided they were pieces of "float" from a large and potentially valuable vein of natural iron buried underground. The cook staked a mining

claim on the area—at the time that was as simple as stacking up a
few rocks to mark the claim—and then took a 40-pound sample
of the iron to Albuquerque, New Mexico, in the hopes of selling
the claim to a mining company. There is no indication that he cut
the Mexican-American sheepherders in on the deal. Armed with
a small advance payment, the cook promptly went on a three-day
drunk and disappeared, but fortunately the Coon Mountain iron
sample he had brought with him remained behind.

And thus through ignorance, greed and drunkenness was our
asteroid of fifty thousand years ago first introduced to modern
science. Through a circuitous set of circumstances, a portion of
the cook's iron sample eventually found its way to the Philadelphia
office of Dr. Albert E. Foote, America's most prominent trader
of collectible minerals. Foote immediately recognized the chunk
of nickel-iron as a meteorite, and thus of considerable interest.
During a trip out West, Foote stopped by Coon Mountain, and
with half a dozen hired local men he gathered up more than a
hundred meteorites that were lying on the ground, most of them
small but including one 200-pounder. Later Foote reported his
meteorite findings in an address to the American Association for
the Advancement of Science in Washington, D.C., titled "A New
Locality for Meteoric Iron."

Meanwhile, back at Coon Mountain, Dr. Foote's visit had
caused a sensation. The locals may not have known much about
asteroids and meteorites, but they knew that if such an eminent
mineralogist from back East was interested in those pieces of iron,
then they must be worth something. Chief among them was one
Fred W. Volz, a German immigrant who ran an Indian trading
post in the nearby community of Canyon Diablo, a small, dusty
railroad flag stop perched on the rim of a deep serpentine chasm
that splits the high desert thirty miles east of Flagstaff. No sooner
had Foote departed the area than Volz went into the meteorite

business, sending out local Navajos and hired hands to search for the "irons." It was a meteorite round-up, with cowboys and sheepherders loping their horses over the plains around Coon Mountain in search of the mysterious space rocks. And they were finding the irons by the wagonload, some pieces weighing up to a thousand pounds.

Eventually Volz and his meteorite herders collected more than 20,000 pounds of meteorites, most of which Volz sold for an average of $1.25 a pound to museums and collectors around the world. Because meteorites are usually named for the town closest to where they are found, the meteorites collected by Volz and others came to be known as Canyon Diablo meteorites—and soon "Canyon Diablos" were featured in every important meteorite collection. The Field Museum in Chicago, the Museums of Natural History in Paris and London, the Smithsonian, the American Museum of Natural History in New York—they all acquired large chunks of our little asteroid of fifty thousand years ago. Canyon Diablos are still popular items in today's meteorite market, with small pieces going for about thirty or forty bucks an ounce, while bigger pieces can run into the thousands of dollars.

Unfortunately for Volz, though, the meteorite market in the 1890s wasn't quite as broad and brisk as it is today. After he disposed of some 10,000 pounds of Canyon Diablos, sales started dropping off, even at a reduced price of fifty cents a pound. The market was glutted, and thousands of pounds of Canyon Diablos wound up stored in barrels at Volz's trading post.

Still, the Canyon Diablo meteorites excited a lot of interest, not only among curators and collectors but among some prominent scientists as well. And in the autumn of 1891 one of those scientists came calling. His name was G. K. Gilbert, and he was the chief geologist for the U.S. Geological Survey, the government's primary scientific research agency.

It's hard to overstate the prominence and influence of Grove Karl Gilbert in American science during the late nineteenth and early twentieth centuries. A lot of scientists are respected, and Gilbert surely was. But he was also actively revered. One historian has described him as "perhaps the closest equivalent to a saint that American science has yet produced." Tall and lanky, with an impressive full beard—in his younger days he looked like one of the guys in ZZ Top—he was an inspiration and mentor to a generation of young scientists within and without the USGS. In fact, maybe "saint" isn't a strong enough word. To many scientists of the day, G. K. Gilbert was more like a god.

Gilbert's interest in Coon Mountain was serendipitous. When Dr. Foote delivered his report on the Canyon Diablo meteorites to the American Association for the Advancement of Science in Washington, Gilbert happened to be in the audience. During his speech Dr. Foote mentioned that the Canyon Diablo meteorites had been found scattered around an enormous crater in the ground, and while Foote didn't connect the two, Gilbert had another thought. As Gilbert later wrote, "When Dr. Foote described a crater in association with iron masses from outer space, it at once occurred to me . . . that [perhaps] the shower of falling iron masses included one larger than the rest, and that this greater mass, by the violence of its collision, produced the crater."

Gilbert had nailed it, sight unseen. That was exactly what happened—although Gilbert was too good a scientist to think of it as anything more than a theory until it could be proved, or disproved. And it was a bold, even radical theory by the standards of the day. Once again, almost no reputable scientists of Gilbert's time would have even considered the idea that an asteroid big enough to blast out a crater a mile wide could possibly have hit the Earth. For a scientist of Gilbert's standing to do so was unusual indeed.

Gilbert eventually mounted an expedition to the Arizona crater, telling a friend, "The errand is a peculiar one. I'm going to hunt for a star." The eminent geologist and his crew spent more than two weeks at the crater, collecting meteorites, making sketches, conducting magnetic tests. Gilbert assumed that if a metallic asteroid had blasted out the giant hole, then the asteroid must still be there, buried underground—in which case his magnetized instruments should be going wild. But they weren't; they showed no evidence of a giant mass of buried metal. (As we'll see, there was actually a very good reason for that lack of magnetic response, but Gilbert had no way of knowing that. The science involved wouldn't be understood until decades later.) The absence of magnetic anomalies, combined with other tests, convinced Gilbert that the crater had not been formed by a plummeting asteroid. Instead, he decided that the crater had been formed by an ancient volcanic steam explosion, and thus was of little scientific interest. As for the metallic meteorites scattered about the area, that was just a coincidence; they must have fallen after the crater was formed.

"I did not find the star, of course," Gilbert somewhat disappointedly wrote his friend. "Because she is not there."

And that was that. As far as the scientific community was concerned, the subject was closed. That mysterious crater in the Arizona desert had been created by good, sound, explicable geological processes, not by some rogue asteroid hurtling down from space. Saint Gilbert had said so.

In retrospect, it was one of the great missed opportunities in scientific history. If Gilbert had looked a little deeper, he could have been the first scientist in history to prove that a large asteroid had struck the Earth, with all the implications thereof. Given his scientific reputation and clout, it could have opened up an entirely new and unprecedented line of scientific inquiry into the Earth's astrogeologic history and the nature of our Solar System. It might even have earned Gilbert one of those coveted "Father of . . ." titles—in this case, "The Father of Meteoritic Impact Theory."

But Grove Karl Gilbert just couldn't make the leap. And for the rest of his days he never publicly mentioned the Coon Mountain crater again.

And there the matter stood. The notion that the great crater in the desert had been created by a falling asteroid persisted among the local Arizona populace, but to the scientific world that was just a legend, a tall tale swallowed only by the gullible. Only a non-scientist would believe such a thing.

Which brings us back to Daniel Barringer.

Although Barringer was previously unaware of Gilbert's ultimate dismissal of the crater as a volcanic structure, not an asteroid impact, he had the same reaction the great geologist initially had. Barringer was certainly aware of meteorites and asteroids. And like Gilbert, Barringer assumed that if a plummeting asteroid had in fact blasted out the crater and scattered nickel-iron meteorites around the area, then the main mass of that giant metal asteroid must also be pure nickel-iron—and it also must still be there, buried under the crater, just waiting to be found.

Unlike Gilbert, though, Barringer's interest was not just scientific. As a professional miner, Barringer realized that if the asteroid actually existed, it might be possible to exploit it for the valuable metals it contained. To put it another way, Barringer's idea was to *dig up the asteroid and sell it.*

From an engineering and mining standpoint, this made sense. After all, if you shoot a rifle bullet into the ground, you can dig down a few inches and there's your bullet, relatively intact. At first glance the same principle should apply if your "bullet" is a giant asteroid. Based on the reported mile-wide size of the crater, Barringer estimated that the mass of the buried asteroid would have to be at least a million tons, maybe more, of pure nickel-iron. It was a miner's dream.

And the dream got even better. Barringer eventually learned that the meteorites found around the crater also contained small percentages of platinum and iridium, which are rare in the Earth's crust but relatively abundant in asteroids. Those two hard, silvery, highly valuable minerals were widely used for everything from dental fillings to jewelry to nib tips on fountain pens, as well as for industrial applications. The upshot was that every ton of the buried asteroid could contain some $200 worth of nickel, platinum, and iridium—and if you multiply that by a million tons of asteroid you get $200 million. And that's in 1902 dollars; today we'd be talking about billions. Barringer was convinced that digging up that asteroid would be one of the most lucrative mining operations in the history of the world.

As for the esteemed G. K. Gilbert's conclusion that the asteroid didn't exist, Barringer dismissed it out of hand. Since when did anyone else's opinion hold any sway over D. M. Barringer?

"I *knew* that the crater was produced by the impact of a great body that had collided with the Earth," Barringer later recalled. "And I *knew* that the mass which made it [lay] at the bottom of it." The asteroid was there. It had to be.

Soon after his epiphany on the hotel porch, Barringer rounded up a few investors and formed a corporation called the Standard Mining Co. He also hired Holsinger away from the General Land Office to oversee the crater mining operation—"If there should be a future in digging up meteorites, you can count me in," Holsinger told him—and then they quietly filed a mining claim on Coon Mountain.

(Coon Mountain was situated on unused federally owned public land, which meant that anyone could file a mining claim on it—a claim that gave the holder the right to extract minerals from it for as long as they lasted, but retained federal ownership of the property. But Barringer was thinking longer term. Filing the claim was the first step in obtaining a mining *patent* on the property, which would give him and his heirs outright ownership of the land in perpetuity—which is why Meteor Crater is still privately owned today. Barringer was extremely secretive during the claim-filing process, fearful that his well-known name might inspire competing claims or even physical claim-jumping. In fact, he often communicated with Holsinger using the Western Union Telegraphic Code of 1903, which listed tens of thousands of nonsense words that corresponded with numbered phrases that the telegram recipient would decipher with his own copy of the codebook. Here's a sample of one of Barringer's telegrams: "*Dorpsland etrurian enoiseler emuscation ippotamia exannulate inesorato chupetilla irrigateur.*" I'm not a cryptologist, but after an excruciatingly tedious study of the codebook, the telegram above appears to decipher as: "Did you find iron in the hole? Reply fully by telegram giving all details. How is the land situated? What will smelting works cost? Is everything covered? Advise by telegraph at earliest opportunity.")

Barringer's asteroid-mining plan was simple enough. First they would dig down through a few hundred feet of accumulated sediment to get to the buried mass of the asteroid, with the spoil cast aside on the crater floor. Then hundreds of thousands of tons of asteroid ore would have to be dug up and hauled out of the crater on a road or rail system to an on-site smelter, where it would be roasted and bathed in sulfuric acid to leach out the valuable nickel and platinum-group minerals. Sure, it would be messy, essentially turning the crater and its environs into a vast, smoky, chemically poisoned open-pit mine. But as Barringer put it, "If that stuff [the meteorite ore] is worth $200 a ton I will make the interior of the crater look like a rabbit warren before I give up my search. . . ."

Well, by this point you may be thinking: This guy was going to do *what*? He was going to deface or even destroy an astonishing natural and scientific wonder—for *profit*? And nobody was going to *stop* him?

Obviously, that's the modern view. And if a reincarnated Daniel Barringer were to announce such a scheme today you can imagine the reaction. He'd be bombarded with lawsuits, howls of social media outrage from around the world, environmental activists chaining themselves to the crater gates. Actually it probably wouldn't even get that far. Two seconds after he presented his crater mining plan to the EPA, Barringer would probably be frog-marched out of the building with orders to never come back.

But we have to look at it in the context of the times. If Barringer didn't seem concerned about the destructive environmental effects of mining the crater, neither did anyone else. I've read scores of early-twentieth-century newspaper stories, magazine articles, and scientific reports about Barringer's asteroid-mining project, and I haven't found a single expression of outrage or even mild concern that this natural wonder was in danger of being drilled, bored, blasted, excavated and smelted for profit. Not one. No one seemed to care. It's been written that "The past is a foreign country: they do things differently there"—and in the foreign country that was early-twentieth-century America, attitudes weren't what they are now. The land and the riches buried underneath it were there to be exploited for man's benefit. And let's be honest. Without mines and miners like Barringer we'd have no civilization. We'd all be tilling the soil with pointed sticks and using clam shells for cutlery.

The point is that Barringer faced no environmental or social challenges to his asteroid-mining plan. And after his application for a mining patent on the crater was approved by the federal Surveyor General's office—there was some discussion about a bribe being involved—there were no potential legal challenges either. Barringer could do with his land whatever he wanted.

And so in the spring of 1904, Daniel Barringer and his small crew of hired miners began to dig.

❍

"A Philadelphia company is digging for a meteor out south of Canyon Diablo," the *Coconino Sun* newspaper in nearby Flagstaff reported that summer. "There are those who believe that an enormous meteorite once struck at this point . . . which, if found, is to constitute a bonanza iron mine."

That was a pretty big "if." Because finding the asteroid bonanza was turning out to be a lot harder than Barringer had thought.

The first effort involved digging a six-foot-square vertical shaft in the exact center of the crater floor. Working in shifts, two guys with picks and shovels would chip away at the bottom of the shaft, loading the freed dirt and rock into a bucket that was hauled to the surface by a workhorse walking 'round and 'round a capstan device called a "horse whim." It was hard, dirty, dangerous work. The operation was plagued by massive thunderstorms and almost constant winds of "near hurricane" strength. The miners were especially afflicted by the clouds of pulverized white silica that the asteroid's tremendous impact had left behind by the millions of tons. The talcum-powder-like silica—Barringer called it "rock flour"—burned the eyes and choked the lungs. Although Barringer was paying his workers a better-than-average wage—$2 a day plus board, with food prepared by a "China cook"—the miners routinely threw down their shovels and quit.

Still, every day Barringer expected the deepening shaft to ring out with the clang of a pickax hitting a million-ton metal asteroid. Unfortunately, at 180 feet the miners hit water mixed with the silica dust, which gave it the consistency of quicksand, and it filled the bottom of the shaft faster than the poor horse could whim it out. So much for that.

Next Barringer brought in a steam-powered "churn drill" to dig a series of small-diameter exploratory holes at various places on the crater floor. The drill dug down three hundred feet—no asteroid. Four hundred feet—no asteroid. A thousand feet—you guessed it, no asteroid. Drills got stuck, bits broke, machinery got clogged with dust. Locally the asteroid-mining scheme was being viewed as a joke.

"We are regarded here as cranks who have no better place to put our money than in that big hole," Holsinger reported to Barringer. "Even the workmen laugh at our supposed folly when our backs are turned."

Barringer was finding plenty of evidence to support the theory that an asteroid had blasted out the crater. The drills brought up samples that tested positive for nickel, a good indicator of meteoric impact. There were also the millions of tons of "rock flour" that had been created by a violent impact. And course there were the meteorites found outside the crater.

But there was still no million-ton asteroid—and the operation was hemorrhaging money. Over two years the digging and drilling cost some $85,000—about $2 million today—much of it Barringer's own funds. Barringer needed fresh money from new investors, but he knew those investors couldn't be drawn in just on his hunch that there was an asteroid down there somewhere. He would need a scientific stamp of approval for the asteroid's existence.

So in 1906 Barringer presented a lengthy paper on the Arizona crater to the Academy of Natural Sciences in Philadelphia. In the paper he offered up the wealth of evidence for meteoric impact mentioned above, and concluded that "We can now prove that this crater is due to the collision with the Earth of an extra-terrestrial body . . . a meteor which must have been of great size."

So far, so good. Barringer laid out his evidence in a calm, intelligent and convincing manner. He didn't get everything right. For example, he estimated the age of the crater at a mere seven hundred to two thousand years, which was about forty-eight thousand years off the mark. Still, for any objective observer the scientific conclusion should have been incontrovertible: A plummeting asteroid had blasted out that giant crater in the Arizona desert. Barringer's paper should have sent a stampede of geologists and astronomers scurrying to the crater to study this amazing scientific wonder. It should have set off a wave of new scientific research into asteroids and meteorites and the collision of celestial bodies.

But Barringer being Barringer, he couldn't leave it with simply proving his case. It wasn't enough that he was right. His adversaries also had to be wrong. Referring to G. K. Gilbert by name, Barringer mockingly ripped to shreds the great man's theory of a volcanic steam explosion, declaring, "it does not seem possible to me that any experienced geologist could have arrived at such a conclusion." There was more about Gilbert, but you get the drift.

Well, by today's low standards of public discourse that may not seem like much of a shot. But in the staid and gentlemanly world of early-twentieth-century science it was a public slap in the face. And you can imagine how that played with members of the scientific establishment, most of whom still revered Gilbert. In the pursuit of commercial profit this . . . this . . . this *amateur* geologist, this glorified *prospector*, was questioning the competence and even the scientific integrity of the saintly Grove Karl Gilbert? They were, almost to a man, outraged.

The scientific establishment didn't respond by publishing their own papers attacking Barringer and his theory. In the world of science, what they did to Barringer was worse. They ignored him.

Even the few scientists who thought Barringer's theory had some merit weren't about to give this arrogant blowhard the satisfaction of hearing them say so. So they acted as if Barringer and his meteorite impact theory didn't even exist. And it drove Daniel Barringer crazy.

Over the ensuing years Barringer continued to argue the case for his buried asteroid, writing countless letters to eminent scientists throughout the U.S. and Europe, and presenting more papers to various scientific organizations. It got to the point where having his theory accepted by the scientific establishment was almost as important as making money off the scheme.

But for years he was met with indifference, or worse, ridicule. For example, during a 1909 address to the National Academy of Sciences at Princeton University, Barringer's alma mater, the audience responded to his asteroid impact theory with audible chuckles and titters of disdain. In another humiliating incident, in 1912 Barringer finally persuaded the American Geographical Society to include a visit to the Arizona crater on its prestigious "Transcontinental Excursion," a grand tour of American natural wonders by a hundred of the world's most eminent scientists. But the tour group only stayed at the crater for a mere two hours before heading off to the Grand Canyon, this after consuming dinner and some forty-eight quarts of beer, a dozen bottles of wine and an unspecified amount of whiskey—all at Barringer's expense. A *New York Times* correspondent with the tour reported that none of the scientists were converted to Barringer's view, and one foreign scientist even accused him of "salting" the crater with meteoritic materials to flimflam them—an accusation that left Barringer apoplectic with rage.

Rounding up financial backers was proving difficult as well. Barringer approached scores of wealthy men, many of them personal friends, wheedling, cajoling, almost begging them to invest in his vision. Barringer's targets usually listened politely—and almost all of them passed, sometimes more than once. You get the feeling that if his rich pals saw Barringer coming down the street they'd duck into a doorway so they wouldn't have to listen to the pitch again. Even Barringer's old friend and now-former president, Teddy Roosevelt, gave him the brush-off. "My dear Barringer," Teddy wrote in 1911. "Do stop to see me in New York anytime you can. . . . When I see you I will explain in more detail the impossibility of my taking an effective [financial] interest in these matters."

(To his credit, Barringer always acknowledged that being able to find the asteroid and profitably mine it was chancy. He also limited his quest for funds to what he called "men of large means, who can afford to take the gamble." Whatever else you want to say about him, Barringer was not a man who would fleece widows and orphans.)

Meanwhile, back at the crater it was the same old story: more drilling, more money, no asteroid. In fact, the only thing that had changed at the crater was its name. In an effort to cut costs, in 1906 Barringer used his political connections to have a U.S. post office established at a railroad flag stop six miles from the crater, with S. J. Holsinger appointed postmaster—a spoils system sinecure whose $100 a month salary was used to offset Holsinger's pay as caretaker at the crater. The new post office was designated "Meteor, Arizona," and the crater eventually came to be known as "Meteor Crater." Again, it was a misnomer; "Meteorite Crater" would have been more technically accurate. Still, it was a lot better than Coon Mountain.

And Meteor Crater wasn't Barringer's only headache during these trying times. His other business interests were suffering as well, eventually forcing him to sell off his Main Line Philadelphia estate and move his family into a series of far less grand rental homes.

"There seems to be no way of diverting the tide of luck which seems to have set steadily against me," Barringer lamented. It almost makes you feel sorry for the guy.

And yet, as ruinously expensive as his asteroid mining scheme was turning out to be, Barringer could find solace in one aspect of his crusade: People were actually starting to believe him.

The shift in attitude began with the general public and the popular press, who were always more willing to embrace a sensational theory than was the scientific establishment. (As we'll see, the same thing would happen with the asteroid impact theory concerning the extinction of the dinosaurs.) No one was interested in a mere geologic sinkhole in the middle of the desert, but a mile-wide crater blasted out by a plummeting asteroid was another thing altogether. Newspapers from New York to Los Angeles printed numerous articles about the "wonderful," "fabulous," "remarkable" Meteor Crater and its celestial origins. Travel guidebooks encouraged train-borne tourists to schedule a stop at Canyon Diablo, advising them that the ubiquitous Fred W. Volz was standing by to take them to see the mysterious crater at $10 a wagonload. Hundreds of tourists visited the so-called meteorite mountain every year, many of them taking away small pieces of meteorites as souvenirs.

(The publicity and the tourism helped popularize Barringer's case, but perhaps predictably, Barringer reacted like an old man yelling at the neighborhood kids to stay off his lawn. "We are strongly opposed to visits by tourists," he complained to a guidebook author. "The amount of valuable meteoric material that tourists have picked up is unbelievable." But the tourists couldn't be stopped. Alternatively, Barringer demanded that the gawkers be charged an exorbitant $3 a head—the equivalent of about $70 today—which makes the current Meteor Crater admission price seem reasonable.)

But it wasn't just tourists and newspaper travel writers who were signing on to Barringer's impact theory. A few prominent scientists—not many, but a few—were coming 'round as well. Notable among them was Swedish chemist Svante Arrhenius, a Nobel Prize recipient who in the 1890s had discovered the "greenhouse effect" caused by man-made carbon-dioxide emissions—what today we call "global warming." Arrhenius reviewed Barringer's work and returned a glowing report, calling Meteor Crater "the most interesting feature on the surface of our planet"—a phrase and source that Barringer used again and again in his investor pitches. (Arrhenius also believed that microscopic life was first carried to Earth by asteroids and comets, a theory known today as "panspermia." As for global warming, Arrhenius, a resident of frosty Sweden, actually thought that a little global warming might not be such a bad idea.)

Armed with growing scientific backing, in 1920 Barringer persuaded the giant U.S. Smelting, Refining and Mining Co. of Boston to finance another round of drilling to find the buried asteroid. It was a disaster, another $175,000 poured down the rathole with no asteroid to show for it.

At this point Barringer was in his sixties, and a lifetime of bad habits was seriously affecting his health. ("I have been poisoning myself eating eggs," he confided to a friend—although, given his drinking and smoking, it's doubtful that eggs were the biggest problem.) But still the man wouldn't give up. Despite all the previous debacles, in 1925 Barringer formed a new company, the Meteor Crater Exploration & Mining Co., with a board of directors and himself as vice president, and miraculously managed to raise yet another $200,000 from investors for more digging.

Ironically, though, the more that reputable scientists came to accept Barringer's asteroid impact theory, and the more they studied the matter, the worse it was for Barringer's asteroid-mining business. Things came to a head in 1929 when the directors of Barringer's new company were becoming increasingly alarmed at the lack of progress in locating the asteroid. So the company directors commissioned an astronomer named Forest Ray Moulton, a renowned expert on celestial mechanics, to take a look at the situation.

And what Moulton came up with was devastating. Moulton proved once and for all that Barringer's million-ton asteroid no longer existed.

Again, it was a matter of kinetic energy. Earlier I used the bullet analogy to describe the asteroid-mining idea: Shoot a bullet into the ground, dig down a few inches and the bullet will be there. Barringer and others actually had shot bullets into rock and dirt to try to prove their case. The problem is that an asteroid plummeting toward Earth at 40,000 miles per hour is not a bullet ambling along at an almost leisurely 2,000 miles per hour. The kinetic energy released by an impacting asteroid weighing hundreds of thousands of tons and traveling at tens of thousands of miles per hour is the energy of a nuclear bomb—and like a bomb, when it goes off it destroys itself.

Moulton's calculations showed that the asteroid was traveling much faster than Barringer had assumed—and as a result the tremendous burst of energy and heat produced by the impact had destroyed almost all of the asteroid. Except for the small pieces of meteoritic material left scattered on the ground, most of Barringer's asteroid had been shattered or melted into billions of tiny pieces that were ejected from the crater and then had blown away with the wind. In other words, Daniel Barringer had spent more than a quarter of a century and most of his fortune chasing a chimera, a metallic will o' the wisp.

Moulton's findings explained why G. K. Gilbert's magnetic tests in the 1890s had failed to indicate a massive metallic asteroid buried underneath the crater; it was because there wasn't one. Gilbert was right in deciding the asteroid did not exist, but wrong in concluding that the crater had not been formed by an impact. Barringer was right in believing an impact had formed the crater, but wrong in insisting the asteroid was still there. Seldom in the history of science have two such forceful and intelligent men been so wrong and so right about the same thing at the same time.

Barringer disputed Moulton's conclusions, of course. "The evidence is all against such hifalutin' theories," he railed. But other astronomers and mathematicians backed Moulton up—and so did the company directors. In September 1929, they shut down any further work at Meteor Crater.

Two months later, at age sixty-nine, Daniel Barringer died of a massive heart attack.

○

As it turned out, Barringer was correct in thinking there is money to be made in mining an asteroid. He was on the right track—he was just looking in the wrong place.

"He certainly was thinking outside of the box," says John S. Lewis, professor emeritus of planetary science at the University of Arizona's Lunar and Planetary Laboratory. "He was dogged, and you can't blame him for his enthusiasm. But the technological knowledge required to assess the scheme simply didn't exist at the time."

Lewis knows something about asteroid mining. As the author of two groundbreaking books—*Mining the Sky*, published in 1997, and 2014's *Asteroid Mining 101*—Lewis is one of a growing number of space pioneers who are developing plans to exploit asteroids for their almost limitless natural resources. The only difference between them and Daniel Barringer is that they obviously aren't going to dig up asteroids from out of the ground. They're going to explore and then mine asteroids while they're circling around in space.

And that's not some wild-eyed futuristic idea, the sort of thing that always seems to hover over the time horizon, like human teleportation ("Beam me up, Scotty") or everyone riding around in flying cars. It's not just the stuff of science-fiction movies like *Alien* or *Outland*. It's happening right now.

"It could be the biggest game-changer in human economic history," says Lewis. "The concept of asteroid mining has engendered the formation of several companies that are attracting venture capitalists. People who want a piece of the action are flocking in."

Lewis is the chief scientist for one of those companies, Deep Space Industries, headquartered in California's Silicon Valley. A frequent partner with NASA in various space projects, DSI is developing a small (100-pound) reconnaissance spacecraft called *Prospector-1* that the company plans to send to a Near-Earth asteroid within the next few years to assess the mining potential. Another asteroid mining company, Planetary Resources, headquartered in Redmond, Washington, has already launched a small "CubeSat" satellite to test the technology for future commercial exploration of asteroids. There are other companies in the field as well.

And it's not just bold entrepreneurs who are interested. NASA is in the asteroid-mining game, too. As I write this, in 2018, an unmanned NASA spacecraft called *OSIRIS-REx*—an acronym tortuously derived from "Origins, Spectral Interpretation, Resource Identification, Security, Regolith Explorer" mission—is on its way to a Near-Earth asteroid named Bennu, with the testing of asteroid-mining feasibility as an important part of its mission. It's scheduled to briefly touch down on the asteroid, collect a few pounds of surface samples, and then return them to Earth in 2023. (More on Bennu and *OSIRIS-REx* later.)

And what valuable commodities do NASA and the private companies hope to get from asteroids? Perhaps surprisingly, the first thing they want to find is water.

Some types of asteroids contain as much as 20 percent water ice by weight that's locked inside clays and other materials. (In fact, the long-ago impacts of asteroids and icy comets may have provided at least some of Earth's initial water supply.) Once extracted, that asteroid water could be used as fuel for spacecraft, either by super-heating it to provide propulsion—sort of like a high-tech steam engine—or by breaking it down into its components, hydrogen and oxygen, for use as rocket fuel.

Large supplies of water would be crucial for any interplanetary manned space missions or space colonies, not only for fuel but also for human consumption and for growing food in space habitats. Water might also be used for protection; water tanks wrapped around manned spacecraft could help shield human occupants from dangerous solar radiation. The idea is that instead of launching space missions from Earth that have to carry their own supplies of fuel or water, which are heavy and thus enormously expensive to launch, space vehicles could launch "light" and then "gas up" in space with fuel or water that's produced from asteroids.

"Asteroid mining is the key to our future expansion into space," says Planetary Resources CEO Chris Lewicki, a former NASA engineer. "It will open a new era of space exploration."

But pumping water out of asteroids isn't what most excites the popular imagination. Most of the buzz around asteroid mining centers on the same materials that Daniel Barringer hoped to find in Meteor Crater—that is, iron, nickel and especially the valuable platinum-group metals (PGMs) like platinum and iridium. Every year a mere 200 tons of platinum worth about $6 billion are mined on Earth. But according to some estimates, the total value of the PGMs contained in all the Near-Earth asteroids combined is a staggering $70,000 trillion. A single mile-wide, metallic Near-Earth asteroid named 3554 Amun—essentially it's a big chunk of stainless steel—has been estimated to contain some $20 trillion worth of metals, including PGMs. There are more valuable rare metals in that one asteroid than have been mined on Earth in all of human existence. And unlike on Earth, the valuable minerals on asteroids aren't just buried deep below the surface. They're right up there on top.

So it's easy, right? Just hop up to little ol' Amun or some other metallic asteroid, start digging out big chunks, bring them back to Earth and all the shareholders get rich. As one overly enthusiastic entrepreneur put it, sounding a lot like Barringer making a pitch to a potential investor, "There are twenty-trillion-dollar checks up there, waiting to be cashed!"

Not so fast, says Lewis.

"People who get bitten by that idea find that it goes sour pretty quickly," Lewis says. "Trying to get rich solely off of platinum-group metals simply wouldn't pay off."

There are several problems with trying to mine asteroids solely for valuable metals that would be used on Earth. For one thing, those fantastic asteroid values mentioned above—$20 trillion, $70,000 trillion—are based on the current Earth price of precious metals. Obviously, if you brought back too much platinum or other precious metals from an asteroid it would collapse the market price. It would be like De Beers finding a million-ton diamond in South Africa and cutting it up and putting it on the market all at once; suddenly your fiancée would be demanding a sapphire or emerald engagement ring, not some worthless old diamond.

Also, to concentrate solely on extracting PGMs from iron-bearing ore on an asteroid would be economically inefficient, requiring massive processing facilities and high transportation costs, and the vast majority of the metal that's processed—the iron—would be wasted. Earth currently has sufficient supplies of relatively cheap iron and other metals, at least for the near future; there wouldn't be any point or profit in bringing it in from space.

The answer, Lewis and others say, is to create a market *in space* for asteroid metals and other materials, such as silicates for glass and ceramics, or carbon and phosphorous as fertilizer for growing food on space bodies or habitats. Say a government or a private company wants to build a 450-ton space station. The cost of sending Earth-produced metals into space to build the space station would be huge, many millions of dollars per ton. But what if you used metals from an asteroid to build your space station?

Steel produced in space from asteroids could be delivered to the space station assembly area at a fraction of the cost of Earth-produced metal. Selling large quantities of asteroid-produced steel for space stations, solar power satellites, modular space habitats and other space construction projects would pay the expenses of your asteroid-mining machinery, which could be operated either robotically or under human supervision. Meanwhile, the iron processing would produce valuable platinum-group metals as a by-product—Lewis calls it "the gravy"—that you could sell in small, non-market-collapsing quantities on Earth to further increase profits. You could also manufacture components or finished products in space—cell phones, for example—using raw materials that are far more abundant in space than on Earth, and then transport the finished product back to Earth for sale at the Verizon store.

Making a profit is the key, asteroid-mining advocates say. Sure, governments can afford to spend a billion here and a billion there on space exploration for purely scientific purposes, but government projects are inherently slow-moving, wasteful and economically inefficient. It's the nature of the beast. But the profit motive will spur innovation, reduce inefficiencies and lower costs. Already private companies like SpaceX have lowered the cost of space vehicle launches. The same would be true with asteroid exploitation.

Is asteroid mining financially risky? Sure. If you're an ordinary wage earner no one is suggesting that you park your 401(k) in an asteroid-mining company. As the Wall Street firm Goldman Sachs said in a 2018 report, "Asteroid mining could very quickly supply an emerging on-orbit manufacturing economy with nearly all the raw materials needed"—although it cautiously described profitable asteroid mining not as a sure thing but as "not outside the realm of possibility." It's been said that asteroid mining in 2018 is the Internet in 1986—that is, for people willing and able to take the risk, it could be the equivalent of buying Amazon stock at eighteen bucks a share.

Actually there's a certain irony at work in all of this. The most easily accessible candidates for asteroid-mining operations are the ones whose orbits bring them closest to Earth—that is, the aforementioned Near-Earth Objects. But some of those financially valuable NEOs also pose the greatest danger of someday colliding with Earth. The joke in the asteroid-mining community is that "Earth is being threatened by huge piles of money!" On the other hand, some of the same techniques and technologies that would be used for asteroid-mining operations could also be used to deflect asteroids heading for Earth—about which more later.

There are some legal and political hurdles to be cleared. For example, the "Outer Space Treaty" of 1967 specifies that no nation can own a celestial body such as an asteroid, but it's a little unclear where international law stands on private exploitation of asteroids. The U.S. government tried to clarify its position with the "Spurring Private Aerospace Competitiveness and Entrepreneurship (SPACE) Act of 2015," which explicitly grants American citizens the right to exploit asteroid resources, including water and minerals, as long as there's no biological life involved. In short, no pushing around the natives like the Resources Development Administration did when it was mining for unobtainium in the film *Avatar*—although we're pretty sure that indigenous peoples are not going to be an issue with asteroids. No evidence of current or past biological life has ever been found on an asteroid or meteorite. (By the way, the director of *Avatar*, James Cameron, was one of the initial backers of the Planetary Resources company.)

There are also some environmental concerns, at least among some people, that asteroid mining would diminish the asteroid population and pollute space with the same toxic waste products produced by mining operations on Earth. A 2018 Pew Research poll found that a slight majority of Americans—51 percent—had little confidence that private companies would try to minimize space pollution. So when and if actual mining operations begin, we can probably expect some sort of "Save the Asteroids!" movement. But realistically, there are millions of asteroids out there, and space is a pretty big place. So if we have to choose, which is better? Producing relatively small quantities of mining pollutants in the vastness of space, or dumping them into our already-compromised ground, water, and atmosphere?

And how long will it be before humans actually start mining asteroids? When Planetary Resources and Deep Space Industries were formed in 2009 and 2013, respectively, there were some predictions that actual mining—for water at least—could begin as early as 2020. Obviously that was overly optimistic. The best current estimates are that it will take at least a decade, maybe longer.

Still, the bottom line is that if it's done correctly, someday asteroid mining can be both economically viable and technologically feasible. And I'm guessing that if he were alive today, D. M. Barringer would be up to his neck in the asteroid-mining business.

☾

All told, Daniel Barringer spent at least a million dollars searching for his asteroid, about half of it his own money. After his death his badly diminished estate was valued at about $20,000, or about $250,000 today—not a pittance, but hardly a legacy fortune for his wife and eight grown children.

Although Barringer's asteroid-mining scheme died with him, his company's directors tried various other ideas to make Meteor Crater pay. There was a plan to package the fine-grained silica powder that had so afflicted the miners and market it as a household scouring powder, like Ajax or Comet; they were going to call it "Star Dust." Another plan was to mine the silica in vast quantities and sell it to glass manufacturers. Fortunately for the preservation of Meteor Crater, neither scheme played out. By the 1950s the Barringer family had dedicated Meteor Crater solely to tourism and scientific research. The only evidence that remains of Barringer's asteroid-mining operation is some old pieces of rusting equipment and a boarded-over mine shaft in the center of Meteor Crater.

As for Barringer himself, he had always been worried—sometimes to the point of paranoia—that he somehow would be robbed of the credit for being the first to prove that an asteroid had struck the Earth. So he might have been posthumously relieved if he could have seen the headline on his obituary in the *New York Times*: "D.M. BARRINGER, GEOLOGIST, DEAD: Discovered Origin of Famous Meteor Crater."

Sadly, outside the scientific community Daniel Barringer is pretty much forgotten these days. Even the Arizona crater into which he poured so much of his life and fortune doesn't bear his name. Some scientists unofficially call it "Barringer Meteorite Crater," but the official name remains Meteor Crater.

In fact, it wasn't until 1971, more than four decades after his death, that the International Astronomical Union finally named an asteroid impact crater after Daniel Barringer. Given his relative obscurity, perhaps it's not surprising that the 15-mile-wide hole named "Barringer Crater" is situated where no one standing on Earth can ever see it—on the far side of the Moon.

Ultimately, did Barringer's relentless crusade to prove that an asteroid impact gouged out Meteor Crater really matter? Absolutely. Before Barringer, almost no one in the science establishment even considered the idea that our planet had been bombarded by asteroids and comets. That changed after Barringer and the scientific discoveries he prompted. It's been said that seeing is believing, but sometimes you to have to believe before you can see. And once they were forced to believe that an asteroid impact was possible, once they knew from Meteor Crater what to look for—meteoric minerals, crushed surface rocks, and so on—other scientists were finally able to see evidence of numerous cosmic bombardments of Earth. And as we will see in the next chapter, not all of those massive bombardments happened in the distant past.

Some happened almost yesterday.

CHAPTER 4

STAR WOUNDS

In the summer of 1908 people noticed a curious phenomenon in the nighttime skies over northern Europe:

They were *glowing*.

It was weird. The skies were lighted up in red and green and golden hues, so brightly that even at midnight it was light enough to read a newspaper outside, or take photographs without a flash lamp. It was as if the sky were in flames—so much so that some people in London alerted their local fire brigades.

And the situation got stranger. Shortly before the heavens lighted up, seismograph machines around the world picked up powerful tremors in the earth, and microbarographs detected a pulse of atmospheric pressure that had traveled thrice around the globe. The epicenter of these odd phenomena seemed to be located somewhere in the broad empty reaches of Russian Siberia.

Everyone knew that *something* had happened out there—but what? An earthquake? An erupting volcano? Some scientists thought the bright skies and other effects were reminiscent of the 1883 volcanic eruption of the island of Krakatoa near Java, an event that blasted at least four cubic miles of rock and ash into the air, much of it in the form of dust particles and gases that lingered in the stratosphere for several years, altering weather patterns and creating fiery sunsets around the world.

News didn't travel fast in Siberia in 1908, but eventually word started leaking out. Russian newspapers reported that on the morning of June 30 a fireball brighter than the Sun had streaked northwestward across the sky before exploding over the virtually unpopulated taiga forests somewhere near the Stony Tunguska River. The fireball created a trail of brilliant light five hundred miles long and set off thunderous explosions that were heard hundreds of miles away. Isolated Russian hunters and reindeer-herding Tungus tribesmen were knocked off their feet by the shock waves and flash-burned by the heat—at least two of them reportedly died, maybe more—and hundreds of reindeer were said to have been killed. In an account eerily similar to Thomas Dekker's description of events in England in 1628, a local Russian newspaper reported that "a loud knocking was heard, as if artillery was fired. . . . All villagers were stricken with panic and took to the streets, and women cried, thinking it was the end of the world."

No one realized it at the time, but what those Siberian villagers and reindeer herders experienced was the most colossal cosmic assault on Earth in modern times. It came to be known as the "Tunguska Event."

Ordinarily such an unusual occurrence would have sent scientists scurrying to the site to investigate. But the epicenter of the explosion was almost impossibly remote, in a roadless expanse of forests and peat bogs. And Russia had other problems to deal with: political unrest against the Romanov regime, World War I, the Bolshevik Revolution. So it wasn't until nineteen years after the event that the Soviet Academy of Sciences finally managed to mount an expedition to the site—an expedition headed by a 44-year-old mineralogist named Leonid Alekseyevich Kulik.

Kulik was the right guy for the job; the man had grit. He had fought in the Tsar's army during Russia's disastrous war with Japan in 1904, got tossed into jail for anti-Tsarist activities, served Russia again as an army officer in World War I and managed to survive the revolution and the ensuing civil war, which millions of other Russians did not. In the early 1920s Kulik was working as director of the prestigious Mineralogical Museum in St. Petersburg, and had become a leading expert on meteorites. Based on earlier eyewitness reports, Kulik was convinced that the Tunguska explosion had been caused by an extremely large meteorite—that is, a small asteroid—that had smashed into Earth.

In a country wracked by war, famine, and economic turmoil, Kulik's theory had both scientific and financial appeal. Russian scientists had heard reports about an American mining company that was trying to dig up a buried metal asteroid in a remote place called Arizona, an asteroid reportedly worth millions of dollars. The feeling was if those greedy capitalists could do it, why couldn't patriotic Russians find and dig up their own multimillion-ruble asteroid for the glory and financial benefit of the Union of Soviet Socialist Republics?

So in 1927 Kulik and an assistant finally headed out to Siberia. It was an arduous trip. The local Tungus tribesmen initially refused to guide Kulik to the site, on the understandable grounds that their angry and vindictive fire-god Ogdy had sent the 1908 explosion as a warning, and they didn't want to piss him off again. Once some guides were sufficiently bribed, Kulik and crew still faced many miles of swollen rivers and impenetrable forests and bogs, made all the more uncomfortable by attacks of voracious mosquitos and incipient scurvy.

But eventually Kulik climbed up a low mountain and looked out at the surrounding taiga—and he could hardly believe what he saw. It was a hundred-square-mile swath of utter devastation, with millions upon millions of large trees—pine, cedar, larch—charred and snapped off at the bottoms of their trunks like matchsticks. The blackened fallen tree trunks all pointed back to a central point, like spokes on a wheel, indicating that they had all been instantaneously knocked down by the shock wave of a tremendous fiery explosion.

Today we're all familiar with scenes of widespread explosive destruction: Hiroshima and Nagasaki, Berlin and Dresden and so many others. But in 1927 nuclear weapons and the carpet-bombing of cities were still in the future. Kulik had never seen anything like this—and neither had anyone else.

The destruction "exceeded all the tales of eyewitnesses and my wildest expectations," Kulik later wrote. And when Kulik finally returned to St. Petersburg, his photographs and newsreel footage astonished the world.

"World's Greatest Meteor: Valley Blasted by Its Fall," the *Guardian* of Manchester reported. "Kulik Returning from Siberia Quest," a *New York Times* headline said. "Investigated Giant Meteor. 100 Square Miles Blasted. Meteoric Iron Deposits Estimated at $1,000,000."

The part about the million-dollar meteorite was premature. For reasons I'll explain in a moment, Kulik didn't find any meteorites of any size at the site, much less million-dollar ones. Nor did he find a large crater at the epicenter of the explosion, as he had expected. But for the first time he delivered in words and pictures the proof of just how destructive a meteoric impact could be, and that such impacts weren't confined to the distant past—with all the ramifications thereof. As one Moscow newspaper put it, "How can we know that another [meteor] will not strike Moscow, London or New York? The Siberian area was completely devastated of all life for an area of 100 square miles. In a densely populated region such a phenomenon would be one of the most appalling catastrophes in human history."

Exactly. If the Tunguska Event had instead been, say, the St. Petersburg Event, it would have killed hundreds of thousands of people and possibly changed the course of twentieth-century history. That the "meteor" didn't hit St. Petersburg or Helsinki or some other city was partly luck and partly a matter of demography. Incoming asteroids obviously hit where they hit; they aren't attracted to any particular region of the Earth. And since cities occupy only a tiny portion of the Earth's surface, the odds of the Tunguska explosion occurring over a city were correspondingly small. Still, it made people think.

Like Daniel Barringer, Kulik was determined, dogged and utterly convinced that he was right. Over the next decade Kulik mounted three more difficult expeditions to the Siberian explosion site. He still didn't find any meteorites, or a massive crater, but he did find a few round boggy depressions, some of them a hundred yards across, which led him to conclude that the incoming asteroid had split into pieces just before impact. Kulik dug and drilled and conducted scientific experiments to find his buried meteorites—and again like Barringer, he failed in every attempt. He would have kept going, except that in 1941, after the Germans invaded Russia, Kulik once again picked up a rifle for the Motherland and at the militarily ripe age of fifty-eight he joined the People's Militia. Unfortunately he caught a bullet in the leg in a battle outside Moscow, was captured by the Germans and died of typhus a few months later in a Nazi POW camp. Like I said, the man had grit.

In the decades after Kulik's death there were numerous scientific expeditions to the Tunguska site, although until the end of the Cold War they were almost all restricted to Soviet scientists. Some of them found traces of meteoric material, but no actual meteorites, and no massive crater in the ground.

So what caused the Tunguska explosion? It's simple enough. On June 30, 1908, a rocky asteroid or loosely packed comet nucleus anywhere from 200 to 600 feet wide slammed into Earth's atmosphere at a speed of about eight miles per second. (Opinions differ, but it probably was an asteroid.) The passage through the increasingly dense atmosphere created tremendous heat and pressure, which ultimately caused the asteroid to blow up into millions of tiny pieces while it was at least three miles above the ground. An airburst of that sort wouldn't gouge out a crater, but the fiery blast scorched and knocked down the millions of trees, while the tiniest particles and gases from the explosion lingered in the atmosphere and caused the brilliant nocturnal light shows seen in northern Europe. Meanwhile, any larger meteorite pieces that reached the ground would have quickly deteriorated or been covered up in the boggy landscape in the almost two decades before Kulik got there—which is why Kulik never found any meteorites.

As for the power of the explosion, it's been estimated at up to fifteen megatons of TNT—about the same as our Arizona asteroid of fifty thousand years ago. The fact that the Arizona asteroid was dense nickel-iron allowed it to survive the atmospheric gauntlet and hit the ground, which left a massive crater but limited the most extensive damage to about a five-mile radius. The above-ground explosion of the much less dense Tunguska asteroid didn't leave a crater, but the destructive power of the heat and blast waves was spread over a much larger area. (The atomic bomb at Hiroshima was detonated about a half mile above that city to accomplish just that—more widespread destruction.) The point is that an asteroid doesn't actually have to hit the ground to be a killer.

And how often does an asteroid airburst explosion of similar power occur? Some estimates put it at about once every couple hundred years, others about once a century—but again, that's a probability, not a prediction. An object the size of the Tunguska asteroid could come along at any time—and given the relatively small size of the impactor, there's a good chance we wouldn't see it coming.

But if the once-per-century estimate is accurate, it means that over Earth's long history it has been hit by Tunguska-sized cosmic assaults *millions* of times. Even in the relatively short span of human existence, there almost certainly were thousands of similar explosions prior to Tunguska. Most of them probably happened over the oceans or the millions of square miles of unpopulated land areas, where no one ever saw them. But it seems certain that at least some humans previously witnessed explosions of Tunguska scale or larger. And maybe those were the last things those humans ever saw.

Not everyone buys the asteroid or comet origin of the Tunguska Event; there have been other explanations, some of them more or less scientific, others not. One Russian scientist suggested that it was caused by a blob of burning plasma hurled toward Earth by the Sun. Another scientist decided it was simply a natural gas cloud that leaked from the Earth's surface and exploded like a giant gas oven. A few American scientists in the 1960s believed it was caused by the collision of anti-matter with ordinary old run-of-the-mill matter. And in 1973 two scientists from the University of Texas published an article in the journal *Nature* suggesting that the impactor was a "mini-black hole," a piece of space matter perhaps the size of a Ping-Pong ball but so incredibly dense that it weighed billions of tons. (You'd probably have to be a physicist to get your arms around the concept.) According to the Texans, the black hole punched into the ground at Tunguska, bored through the Earth and popped back out somewhere in the North Atlantic; that scenario was quickly shouted down by other physicists.

And then there are the alien theories. In the 1940s and '50s a Soviet science-fiction writer popularized the idea that the Tunguska Event was caused by the explosion of a nuclear-powered Martian spaceship that was trying to get fresh water for its parched planet from nearby Lake Baikal. Another story had it that the explosion over Tunguska happened when a Japanese spaceship trying to get home after a 2,000-year interstellar journey somehow overshot the runway in Tokyo and crashed in Siberia. And then—my personal favorite—there's the tale about the aforementioned eruption of Krakatoa and a star called 61 Cygni. It seems that the 1883 Krakatoa explosion sent an electromagnetic pulse into space, where it was eventually picked up by residents of the binary star system 61 Cygni, some eleven light-years away from Earth. Wanting to be polite, the Cygnians sent a message back by laser beam, but— oops!—they accidentally cranked up the power knob to, like, *eleven*, and the laser beam wound up blasting millions of innocent trees in Siberia. Earth is still awaiting an apology.

All of those tales started off as science-fiction stories, and as such they were harmless. But over the years the science-fiction somehow turned into science-fact, at least in the minds of some people whose belfries apparently aren't entirely free of bats. Tunguska has become a staple in the UFO and conspiracy communities, with various "proofs" offered up to support the alien Tunguska blast theories. For example, in 2004 a Russian "UFO investigative team" went to the still-remote Tunguska site to search for the remains of a crashed spaceship, and—surprise!—they found it, or said they did. Actually, the piece of spaceship they found looked pretty much like an ordinary rock, but it still got big play on the Internet.

With apologies to the believers, to me it seems like a waste of time. There's never been any hard evidence that Earth has been visited by alien beings, but there's plenty of hard evidence that Earth has been visited countless times by big alien rocks. That evidence presents itself in the form of geologic features that have come to be known as "astroblemes," which is ancient Greek for "star wounds."

Meteor Crater in Arizona is a star wound. The Tunguska epicenter was a star wound, too, or at least a temporary one. And beginning in the 1930s, other star wounds on the Earth's surface suddenly started popping up all over the place.

○

In 1932 an English gentleman with a quintessentially English name set out to do a quintessentially English thing: He tried to find a lost city in a trackless wasteland far from home.

The Englishman was Harry St. John B. Philby, a noted explorer associated with the Royal Geographical Society. A contemporary of T. E. "Lawrence of Arabia" Lawrence, Philby was a flamboyant figure who converted to Islam, dressed in flowing Arab robes and sometimes called himself "Sheik Abdullah." At various times he worked as an intelligence officer for the British Colonial Office, as an adviser to the House of Saud and as an international schemer with anti-Zionist, pro-Nazi sympathies. (One of Philby's sons was Kim Philby, the infamous turncoat British spy and Soviet mole from the 1940s until his defection to the Soviet Union in the early 1960s.)

As for the lost city, it was called Wabar, and it was situated in the Saudi Arabian Rub' al Khali, or "Empty Quarter," a vast desert of shifting sands and brutal heat that virtually no Europeans had ever penetrated. According to Arab legend, in retaliation for various crimes and misdemeanors Allah smote the great city of Wabar with fire and fury, leaving behind only ruined walls and a mysterious piece of iron as big as a camel. Some people called the lost city "The Atlantis of the Sands."

The Atlantis of the Sands! For Philby this was irresistible stuff. So he marched into the Empty Quarter with a caravan of fifteen fractious camels and a few reluctant Bedouins as guides, and after many travails—what would an English gentleman's expedition be without travails?—he finally arrived at the suspected site of the ancient metropolis. There was no city. All Philby found were some large circular holes in the ground, the biggest about 400 feet wide, almost completely filled with sand and surrounded by upraised rims. Scattered around the site were pieces of black glass and some pieces of iron, but none was the size of a camel; the biggest was about the size of a rabbit.

Philby figured he had struck out, that the round holes in the ground were just some ancient extinct volcanos. But when he finally got some samples back to London, a mineralogist determined that the small pieces of iron were nickel-iron meteorites, and the black glass was sand that had been subjected to the high heat and pressure of an asteroid impact. And that impact hadn't occurred in the ancient past; later investigation showed that the asteroid impact was probably the end result of a massive fireball that was seen streaking over Saudi Arabia in the mid-nineteenth century. Philby hadn't found his Atlantis of the Sands, but he had found one of the most recently formed large impact craters ever discovered.

And Philby wasn't the only one who was finding astroblemes. In a remote section of central Australia a sheepshearer stumbled onto a round crater some 500 feet wide that also showed evidence of meteoric impact. In the Ashanti region (now Ghana) in Africa, a geologist who was examining the five-mile-wide Lake Bosumtwi suggested that it had been created by a giant stony asteroid that had slammed into Earth perhaps a million years ago. Even before Philby ventured into the Empty Quarter, a geologist noticed a cluster of circular impact craters just outside of Odessa, Texas, the largest of which was 600 feet in diameter. The shallow craters were mostly filled in by sand and other debris—at one point they were used as a county trash dump—but the ground showed signs of an impact by a small metallic asteroid.

Those were just a few examples. From India to Canada to Estonia and around the world, geologists were taking a new look at craters that had long been thought to be old volcanos—and now that they knew what to look for, they were speculating that they had actually been formed by plummeting asteroids. And some of those craters were monsters. The Vredefort crater in South Africa was almost two hundred miles wide, which meant that the asteroid that created it had to be at least six miles in diameter, maybe more. Although badly eroded in the two billion years or so since the impact, the crater's circular shape is still distinctly visible from satellites.

Of course they didn't have satellites at the time. But the rise of air travel in the 1930s and afterward was an important factor in discovering new impact craters, for the obvious reason that things look different from above than from the ground. Stand on the pebbly shore of a big round lake and it will seem as if it's wider than it is across. But fly over it in an airplane and the round shape snaps right out at you. Pilots of the day started seeing more and more almost round depressions on the Earth's surface, some of them lakes, some of them dry holes. For example, in 1947 a pilot for an oil company spotted a perfectly round crater near Wolfe Creek in Western Australia that was almost as big as Meteor Crater—some 3,000 feet wide and 150 feet deep. It was another astrobleme.

And why are meteorite impact craters always round or nearly round? That was a problem that had stumped Gilbert and Barringer and other early crater investigators. Sure, they could understand why a plummeting asteroid coming straight down at a 90-degree angle could make a round depression. But what if the asteroid came in at a low angle? Wouldn't it plow a huge furrow in the ground before coming to rest? They couldn't figure it out.

Not until the aforementioned Forest Ray Moulton and others discovered the explosive nature of an asteroid impact did scientists begin to get a handle on it. Think of it this way: If you hurl a rock at a low angle into a patch of muddy ground, it will make a skid mark before it comes to rest. But now take a U.S. military MK3A2 concussion grenade packed with eight ounces of high explosive and hurl it at the same patch of muddy ground. (Actually, don't try this at home.) The grenade will make an irregularly shaped skid mark, too—until it explodes. The explosion will obliterate the grenade and the initial skid mark, leaving behind an almost perfectly round hole. Same thing with a hypervelocity asteroid that hits the ground and blasts out a crater.

In any event, by the middle of the twentieth century there were dozens of suspected asteroid or comet impact craters that had been identified on Earth. Not all scientists were signing on to the new impact theory, though. Even in the early 1960s a number of reputable geologists still insisted that the so-called impact craters being found on Earth were caused by tried-and-true geological forces such as volcanos, not by hurtling space rocks.

But there was another, richer source of impact craters that probably should have convinced them. It was on the face of the Moon.

○

Today anyone can take a $49.99 telescope from HobbyTown and point it at the Moon and see that it's covered with impact craters caused by collisions with asteroids and comets. We know they are impact craters because we *know* they are impact craters; that's been common knowledge for decades. But it wasn't so simple for our scientific grandfathers and great-grandfathers.

Ever since Galileo, people had been studying the Moon through increasingly powerful telescopes and wondering about those lunar craters, some of which are hundreds of miles wide and several miles deep. By the nineteenth century the scientific consensus was that since the lunar craters *looked* like the volcanic craters found on

Earth, they obviously came from the same source—that is, they were created by volcanic activity early in the Moon's formative years, when it was boiling and bubbling like a pot of hot oatmeal. Case closed.

Actually, well into the twentieth century most professional astronomers didn't even bother to look at the Moon, at least not for scientific purposes. That was partly because the Moon was, well, boring. The mountains and *maria*—Latin for "seas"—on the Moon's visible face had all been thoroughly mapped and measured, as had most of the thousands of craters big enough to be visible from Earth with the telescopes of the day. The attitude was that everything that could be discovered about the Moon had already been discovered—so why waste your time? And there was another factor. Not only was the Moon well-plowed scientific territory, but among professional astronomers it had become something of a joke.

You can blame that on the "man-bats."

Since the dawn of human history people had wondered whether the Moon and other celestial bodies were inhabited, perhaps by creatures similar to ourselves. And in the eighteenth and early nineteenth centuries the popular theory was that yes, they were home to some kinds of creatures. After all, why would God have created all those other worlds only to leave them barren and empty? The theory known as "cosmic pluralism" held that the Moon and all of the stars, planets, and even those recently discovered asteroids—all four of them—must be teeming with life. One popular Scottish author and minister somehow calculated there were precisely 21,891,974,404,480 living inhabitants in the Solar System, including a whopping five billion on the Moon alone.

And it wasn't just theologians who thought so. Distinguished scientists of the era agreed. In 1780 the eminent astronomer William Herschel, the discoverer of Uranus, declared that it was "almost an absolute certainty" that the Moon was inhabited. Carl

Friedrich Gauss, one of the most renowned mathematicians of all time, suggested building giant heliotropes to send light beams up to the Moon to let the "lunarians" know we were here. And in the 1820s a Bavarian urologist-turned-astronomy-professor named Franz von Gruithuisen claimed that through his telescope he had seen man-made structures on the lunar surface. There were numerous others who made the same sorts of claims.

The "men-on-the-moon" idea came to a head in 1835, when the *New York Sun* newspaper ran a series of articles purportedly based on lunar discoveries by the renowned British astronomer Sir John Herschel, William's son. According to the newspaper, Herschel had developed a telescope of new design and enormous power, and through it he had spotted a host of fascinating phenomena on the lunar surface. There were lakes and white sand beaches, forests and cultivated fields, a string of pyramid-shaped structures. There were blue goats, thundering herds of bison-style beasts, and giant two-legged beavers that walked upright and carried their young in their stumpy beaver arms. Most astonishing were certain human-like creatures, furry little things that stood about four feet tall and had large membranous wings. According to the *Sun,* Herschel had dubbed the species "Vespertilio-homo," or "man-bat." The newspaper even managed to work a little sex into the story, delicately noting that the male and female man-bats openly engaged in "amusements [that] would but ill comport with our terrestrial notions of decorum."

Fornicating man-bats! Blue goats! Bipedal beavers! This was extremely hot stuff. Of course it was all made up, the product of a *Sun* reporter's vivid imagination; the astronomer Herschel had nothing to do with it. But by the time the hoax was exposed, the *Sun*'s widely reprinted series, complete with elaborate illustrations, had already fooled millions of people—including, to their embarrassment, some scientists. As a result, "The Great Moon Hoax" helped put the study of the Moon in bad odor for decades. As one nineteenth-century astronomer put it, "Today one directs to astronomers questions about the Moon only in jest."

Actually there were a few late-nineteenth-century scientists who dared to take a stab at the Moon—including our old friend Grove Karl Gilbert of the U.S. Geological Survey. Strangely enough, while deciding that Meteor Crater was not the result of a cosmic impact, Gilbert got the notion that the craters on the Moon *were* the result of cosmic impacts, not volcanos as everyone thought. Gilbert conducted various experiments on impacts, shooting bullets into mud and stone, dropping marbles into bowls of mush, and so on. He also spent a couple of weeks examining the lunar surface through a telescope at the Naval Observatory—an activity that led one U.S. congressman to declare that "So useless has the [Geological] Survey become that one of its most distinguished members has no better way to employ his time than to sit up all night gaping at the Moon."

Gaping at the Moon! It just goes to show you that anti-science sentiment in the political world didn't start with global warming.

Anyway, Gilbert came up with an unwieldy theory that the lunar craters had been created by "moonlets" that in the ancient past had been in orbit around the Earth and had eventually smashed into the Moon. But despite Gilbert's prominence, the theory didn't go anywhere and eventually he dropped his Moon studies altogether. For most professional astronomers and scientists in the first half of the twentieth century, the Moon was still a big, boring joke—and the craters were still volcanic.

That began to change—slowly—in the 1940s. In 1946 a former Army Air Corps pilot and respected geologist named Robert S. Dietz published a paper that persuasively argued that almost all lunar craters had been caused by meteoric impacts; it was Dietz who came up with the term "astroblemes" for impact craters. Later, in 1949, an astrophysicist named Ralph B. Baldwin published a short book, titled *The Face of the Moon*, that also convincingly advanced the impact theory of lunar craters. The scientific evidence was there for anyone who chose to see it.

Unfortunately, most of the scientific establishment still chose not to see it. The lunar craters were created by volcanism—and since no one could get to the Moon, or even close to it, how could that be proven wrong? As for the craters being discovered on Earth, sure, anyone could *claim* that they had been created by impacting space bodies. But really, where was the scientific proof?

Still, the work by Dietz and Baldwin and others inspired members of a new generation of scientists to take a fresh look, not only at the craters on the Moon but also the suspected impact craters on Earth. Which brings us to one of the most important and influential figures in the history of rocks from space: a geologist, astronomer, teacher, and visionary named Eugene Merle Shoemaker.

$$\bigcirc$$

Gene Shoemaker was one of those guys who just about everybody liked and admired. Ask almost anybody who knew him and what you'll hear is "Great guy!" "Wonderful man!" "Terrific person!" Those verbal exclamation points are theirs, not mine. They really mean it.

It's not that Shoemaker was always easy. Even his friends admit that at times he was impatient, demanding, a perfectionist who in moments of frustration could explode in anger; once he got mad at his stalled car and slammed the door so hard that it shattered the window. He had a tendency to work furiously on a project and then suddenly drop it when something else caught his interest—and he was interested in almost everything. He was a terrible administrator, a man who as a university department head would spend hours strolling around campus talking with students while boring but important paperwork piled up on his desk. Even in those days before credit and debit cards he often carried no cash, and sometimes absentmindedly forgot to pay back small lunch-money loans.

In short, Gene Shoemaker could be frustrating, exasperating, maddening—but that was just Gene being Gene. Shoemaker was one of those fortunate souls whose trespasses were almost always forgotten, whose small sins were overwhelmed by his infectious personality and his zest for science, and for life. Armed with a quick smile and a boyish face that even a Clark Gable–style mustache couldn't age—in his mid-thirties he was still getting carded in bars—Shoemaker made friends easily and often, and kept them. Almost no one could stay mad at Gene.

And he was also brilliant.

"Gene Shoemaker was the foremost planetary scientist of the twentieth century," says Clark R. Chapman, himself a world-renowned figure in planetary science. "He was an intellectual giant."

Born in California in 1928, Shoemaker enrolled at the California Institute of Technology (Caltech) at age sixteen and managed to get his master's degree in geology there just five years later; early on, some of his friends started calling him "SuperGene." In 1950 he was hired as a geologist by the USGS, assigned to search for uranium deposits in the American Southwest. Uranium was a high priority for the U.S. government back then, both for making nuclear weapons and for the expected surge in civilian nuclear power plants. It was while he was looking for uranium in northern Arizona that Shoemaker and his new wife, Carolyn, happened to stop by the famous Meteor Crater. (Actually, on that first visit they had to sneak in; they couldn't afford the fifty-cent admission price.) Like so many others, he was awestruck at what he saw.

As a geology student Shoemaker had always more or less accepted the establishment view that geologic formations like Meteor Crater were volcanic, not the result of meteoric impacts; that's what geology students of the day were still being taught, if they were taught about it at all. But after looking at the crater, Shoemaker wasn't so sure. For one thing, he quickly realized that some of the rocks he saw at Meteor Crater could only have been formed by temperatures greater than 3,000 degrees Fahrenheit—hotter than any volcanic material. And there was something else. To him, the big crater that had been blasted out of the Arizona desert looked a lot like what would happen if someone had set off an underground nuclear explosion.

Shoemaker knew about the effects of nuclear bombs. In the early and mid-1950s the U.S. government conducted a series of nuclear bomb tests in the desert outside of Las Vegas, Nevada, tests with code names like Operation Buster, Operation Teapot, and so on. The purpose was not only to test the atomic weapons themselves but to assess the physical and mental effects on troops operating in the "nuclear battlefield." You've probably seen old black-and-white photos of American soldiers staring at a mushroom cloud rising in the distance, after which they conducted maneuvers in the blast area—which was a really bad idea. Many of the GIs suffered long-term health problems caused by radiation exposure. As part of his uranium investigations Shoemaker had studied some of the relatively small (about 300 feet wide) craters made by underground 1.2 kiloton nuclear bomb explosions during the Nevada tests. And as Shoemaker saw it, Meteor Crater was "pretty much just a scaled-up version" of the nuclear bomb craters.

For Shoemaker, Meteor Crater became a kind of obsession; he later wrote his Princeton PhD dissertation on it. With his ever-present geologist's hammer and a magnifying hand lens, he crawled all over the crater, studying the uplifted layers, peering at rocks and collecting evidence. He even had himself dangerously lowered by rope into Barringer's old and crumbling mine shaft to take rock samples. Clearly, this was a geologist with stones.

His efforts paid off. In 1960, Shoemaker and two colleagues discovered that rocks from Meteor Crater contained "coesite," a mineral that had been artificially produced in a lab by subjecting quartz to high temperatures and pressures of 300,000 pounds per square inch—sort of like whacking a piece of rock with the proverbial million-pound hammer. No natural force on or near the Earth's surface could produce that kind of heat and pressure. Only two things could: a nuclear bomb explosion or the mighty wallop of an impacting hypervelocity space body. And since there obviously were no nuclear bombs fifty thousand years ago, the coesite found at Meteor Crater had to have been formed by the enormous force of a giant plummeting rock from space.

This was the scientific proof the cosmic impact community had been waiting for. Now if you had a big hole in the ground, and you found some coesite in it, you pretty much had yourself an asteroid impact site. Suddenly those suspected impact craters people had been finding around the world became *proven* impact craters. Shoemaker himself discovered a big one during a vacation in Germany in 1960, when he noticed that rocks in a 15-mile-wide circular geologic depression surrounding the Bavarian town of Nördlingen looked familiar. The depression had long been assumed to be volcanic in origin, but when tested the rocks revealed—you guessed it—coesite, which had been formed when a space body a mile wide slammed into the Earth some fourteen million years ago. (Today there are about two hundred confirmed impact craters, the largest of them the aforementioned 200-mile-wide Vredefort crater. Of course that's only a tiny fraction of the historic total; again, most of Earth's impact craters have eroded away or been destroyed by tectonic shift.)

The crater discoveries by Shoemaker and a few others should have settled the issue: Earth had indeed been repeatedly pummeled by asteroids and comets, and not always so very long ago, either. But amazingly, many geologists—usually the older ones, stuck in their ways—*still* weren't buying it. Even the discoveries of other impact-proving substances—another high-pressure mineral called "stishovite," and rock formations called "shatter cones"—couldn't convince them that meteoritic impacts had been part of Earth's geologically recent history. What finally brought everyone around wasn't anything that Shoemaker and others found on Earth; it was something they found during the "Space Race."

Readers of a certain age will recall the near panic that swept America after the Soviet Union launched the Sputnik satellite in 1957—the first-ever man-made object sent into Earth orbit. I can still remember the worried look on my first-grade teacher's face as she told us that those godless Communists had beaten America—a pretty scary thing for a bunch of six-year-olds, even if we weren't exactly sure what a Communist was. As for the American people in general, they could hardly believe it. The Russians couldn't even build a decent washing machine, but they put a *satellite* on a rocket? And sent it into *space*? Actually the satellite wasn't much, just a two-feet-wide metal sphere that sent out radio signals that went "Beep! Beep! Beep!" as it sailed over the American heartland. Still, the fact that the Russians could get the thing up there was a shock, especially since at the time the U.S. hardly had a space program to speak of. There was no NASA, and as far as most people in the U.S. government and military were concerned, rockets were for lobbing nuclear bombs, not space exploration.

That all changed after Sputnik. The U.S. launched a crash space program, schools put new emphasis on science and engineering, the newly formed NASA shot Alan Shepard into a sub-orbital flight and put John Glenn into orbit—albeit well behind the

Russians. President John F. Kennedy famously announced in 1961 the American goal of "landing a man on the Moon and returning him safely to the Earth" before the end of the decade.

Gene Shoemaker wanted to be a part of that. As a young geologist he had long dreamed of going to the Moon—and to him it only made sense. After all, he was a geologist, and geologists studied rocks, and the Moon was just a great big rock, right? In 1960 he persuaded the USGS to set up an Astrogeology Research Program—"astrogeology" being the geologic study of the Moon and the planets and other space bodies. It was a brand-new word for a brand-new scientific discipline, and as the program director Shoemaker became the world's first astrogeologist. It also put him in line to become the first scientist-astronaut on the Moon.

Unfortunately, in 1963 Shoemaker was diagnosed with Addison's disease, which, while treatable with cortisone, put a medical end to his astronaut dreams. As Shoemaker later put it, "Not going to the Moon and banging on it with my own hammer [was] my biggest disappointment in life."

Still, while working with NASA Shoemaker played a key role in a series of unmanned lunar missions to map and photograph the Moon's surface at close range. He also worked on the manned Apollo missions that ultimately led to Neil Armstrong's "one small step for a man" in 1969. As part of their training, Shoemaker took the Apollo astronauts to Meteor Crater to teach them what to look for on the lunar surface. You may also remember seeing him on CBS News with "Uncle Walter" Cronkite, providing live commentary during some of the Apollo Moon missions.

Those unmanned and manned missions to the Moon did more than just beat the Russians. They also proved once and for all that cosmic collisions were a fundamental factor in the history of the Solar System. For one thing, they revealed that the lunar surface isn't covered with thousands of impact craters; it's covered with countless impact craters, from the long-known giant ones down to tiny "zap pits" in the lunar rocks caused by tiny bits of hurtling space particles. The Moon's craters have craters that have craters—and so on, virtually ad infinitum. And the craters weren't all remnants of the ancient past; they were still being formed by impacts of space bodies, the smallest ones on a daily basis, larger ones over longer but still relatively recent periods of time. When "Moon rocks" were brought back to Earth for analysis, researchers determined that while the Moon had indeed undergone volcanic activity in its early life, virtually all of the craters had been formed by meteoric impact.

The conclusion was inescapable. The Moon is just 240,000 miles away from Earth. So if the lunar surface had been smacked countless times by impacting asteroids and comets, then the Earth had to have been similarly smacked over the course of its 4.55-billion-year existence. In fact, because the Earth is a bigger target, it actually has been clobbered by even more impacts. If it wasn't for our atmosphere, which burns up the smaller incoming space rocks, and the erosion that destroys the big craters, Earth's surface would look just like the battered face of the Moon.

(By the way, how do we know that the Earth is 4.55 billion years old? It's because in the mid-1950s a geochemist named Clair Patterson subjected meteorites found around Meteor Crater—the Canyon Diablo meteorites—to extensive lead-isotope testing. Based on the Canyon Diablo meteorite data, Patterson set the Earth's age at 4.55 billion years, give or take 70 million years or so, a figure that has stood up ever since. Patterson later went on to save the world from lead poisoning—but that's another story.)

And it wasn't just the Earth and Moon that have been battered. Contemporaneous and subsequent unmanned probes revealed that Mercury, Venus, Mars, and other space bodies had all been marked by asteroid and comet impacts as well. Mars alone has more than 600,000 impact craters bigger than a half-mile across, and countless smaller ones. And as I mentioned in chapter 2, even asteroids have impact craters caused by collisions with other asteroids.

In short, it's a shooting gallery out there.

"It's like being in a hail of bullets going by all the time," Shoemaker said of asteroid impacts and Earth's place in space. "They are bullets, they're bullets out there in space. These things have hit the Earth in the past, they will hit the Earth in the future. It will produce a catastrophe that exceeds all other known natural disasters by a large measure."

And how many of those "bullets" were there? How big were they, and how often did they hit or pass close to Earth? Shoemaker wanted to find out.

So the geologist became an astronomer. Working with a planetary scientist named Eleanor "Glo" Helin—one of the relatively few women of her day to crack the scientific "boys' club"—in 1973 Shoemaker set up the Palomar Planet-Crossing Asteroid Survey, named after the Palomar Observatory, on a 5,600-foot mountain peak outside of San Diego where they did their telescopic observations. The survey's mission was to try to find the asteroids that pose a potential threat to Earth.

This was something completely new; no one had ever conducted a systematic search for potentially dangerous asteroids before. There had been a few efforts by astronomers to study asteroids in the 1950s and '60s, most notably by Gerard Kuiper—he of the Kuiper Belt of comets—and Dutch-American astronomer Tom Gehrels. But as I mentioned earlier, asteroids hadn't exactly been a popular field for scientific inquiry in the first six decades of the twentieth century; it was the "vermin of the sky" thing. As of the early 1970s only a few thousand asteroids had been identified, some by amateur astronomers, others by professionals who were actually looking for other things they thought were more interesting—stars, galaxies, supernovae. And almost all of those known asteroids were safely stored away in seemingly stable orbits in the asteroid main belt, hundreds of millions of miles away; they never come close to Earth.

But some asteroids do come close to Earth—or at least close by space standards. Those are the aforementioned Near-Earth Objects, or NEOs. (The vast majority of NEOs are asteroids or fizzled-out old comets, but a few are active comets that spew out gases and dust as they whiz past the Sun.) By definition an NEO is any object that comes within 30 million miles of Earth's orbit—and if they stayed that far away we wouldn't have anything to worry about. But some NEOs come a lot closer, some of them actually intersecting Earth's orbital path as they pursue their own orbits around the Sun. If they're bigger than 450 feet wide and they come within 4.6 million miles of Earth they're classified as "potentially hazardous objects." (The largest NEO is 1036 Ganymed, a crater-pocked stony asteroid about 22 miles wide. But the vast majority of NEOs are much, much smaller.)

And after they make a close pass by Earth those potentially hazardous NEOs don't just fly off into space, never to be seen again. They come back again and again as they continue to orbit around the Sun. At some point the asteroid may be gravitationally nudged into a different orbit that flings it out of the Solar System or causes it to collide with a planet. But if not, if the asteroid continues to intersect Earth's orbit, eventually Earth and the asteroid will find themselves in the exact same place in space at exactly the same time.

Think of it this way. Imagine a car endlessly driving around a giant circular racetrack; that's Earth orbiting around the Sun. Now superimpose an oval-shaped racetrack over the first one and then send another car driving around it; that's a potentially hazardous asteroid orbiting the Sun. Most of the time the Earth-car and the asteroid-car won't meet at the place where the two racetracks intersect. They may miss each other by a mile or they may miss by inches, but it will still be a miss. But if they both keep going around their tracks long enough, eventually—wham!—they're going to collide. Of course, it's complicated by the fact that Earth and asteroids aren't circling around on the same flat, two-dimensional tracks; instead, the two racetracks are often tilted three-dimensionally in relation to each other. But you get the idea.

Anyway, when Shoemaker and Glo Helin started the Palomar Planet-Crossing Asteroid Survey in 1973, nobody had a clue as to exactly how many Near-Earth asteroids were out there. Only a couple dozen of them had been discovered in the previous seventy years, and many of those had been "lost," meaning their orbits hadn't been calculated and no one knew where they were. Extrapolating from the impact craters on the Moon, Shoemaker estimated there must be at least two thousand Near-Earth Objects bigger than a half-mile across that could someday pose an existential threat to human civilization, along with countless smaller ones that could produce lesser but still catastrophic damage. Shoemaker and Helin planned to find as many of those threats as they could.

Once again, it's not easy to spot asteroids, even the ones that come close to Earth. Asteroids are relatively small, and they don't reflect much light. (In his inimitable style, Bill "The Science Guy" Nye, the bow-tied TV personality and CEO of the Planetary Society, notes that "Looking for an asteroid is like looking for a charcoal briquette in the dark." Founded in 1980 by popular astronomer Carl Sagan and others, the 50,000-member Planetary Society is a private, nonprofit organization that promotes space exploration.) Also, the sky is a big place, and you have to be looking at the right place at the right time. And back in the days before computers were easily accessible, asteroid spotting was a laborious process.

Using an 18-inch "astrophotographic" telescope (18 inches is the size of the aperture, not the length of the telescope), Shoemaker and Helin had to hand cut and load long-exposure film into the telescope, take a 20-minute exposure of a small patch of sky, then take another ten-minute follow-up exposure, then develop the film and tediously go over the images to try to find the dim, blobby smear of an asteroid moving against a stationary backdrop of stars. Then they had to repeat the process again, and again, and again. And they had to make their observations in the confines of a dome-shaped metal structure that was partially open to cutting winds and freezing mountaintop temperatures.

It took Helin and Shoemaker six months to find their first NEO, a mile-wide piece of rock unglamorously named Asteroid 5496 that had passed within eight million miles of Earth—not a particularly close miss, but still putting it well within the Near-Earth neighborhood. In 1976 Helin discovered the first Aten-type asteroid, a group of Earth-crossers that have a high potential for Earth collisions. (Based on the shape of their orbits, NEOs are assigned to several different groups—Apollos, Atens, Atiras, Amors.) But it was slow going. During the first five years the Shoemaker-Helin team discovered plenty of other main belt asteroids, but only a dozen potentially dangerous Earth-crossing asteroids.

Eventually, after some conflicts with the strong-willed Glo Helin—she apparently was one of the few people who could stay mad at Gene—Shoemaker professionally separated from Helin and founded another "sky survey." This was the Palomar Asteroid and Comet Survey, which used the same 18-inch Palomar telescope as the original survey. This time Shoemaker was working with his wife, Carolyn, a mother and former housewife who showed a natural talent for spotting asteroids and comets. Using improved, "hypersensitized" film and a new stereomicroscope device to examine films, over the next few years the husband and wife team discovered scores of Near-Earth asteroids and a dozen new comets.

Unfortunately, at the time not many people in the scientific world seemed all that interested in potentially hazardous space objects. It was still the stars that beckoned; asteroids were a research backwater. The Shoemakers saw their funding cut back, their access to telescope time threatened. When it came to the asteroid threat, they and a few others were lonely voices in the wilderness.

Which was strange, given the fact that in the decades since the Tunguska Event there had been a steady stream of reminders about the potential threat posed by incoming asteroids. In 1930 a Franciscan priest reported that villagers near the Curuçá River in Brazil had been terrorized by a huge fireball that exploded in the sky and covered the ground with red dust and ash. In 1947 a 30-ton metallic asteroid exploded in the atmosphere over the Sikhote-Alin Mountains in far-eastern Russia, scattering meteorites over a large area and creating a series of craters up to a hundred feet wide; it was the biggest known cosmic collision since Tunguska. In 1969 a monster fireball lighted up the ground for hundreds of miles around the small Mexican village of Pueblito de Allende and deposited more than two tons of meteorites. In 1972 an asteroid the size of a Greyhound Scenicruiser blazed through the skies at nine miles per second over Utah and Montana before skipping out of the atmosphere and heading back into space; if it had hit Salt Lake City it could have killed thousands.

There were more to come. In 1992 thousands of people saw a giant fireball streak across the sky over West Virginia and Pennsylvania before it blew up and sent a 26-pound chunk of stone crashing into a parked 1980 Chevy Malibu in Peekskill, New York; it totaled the $300 car, but its owner, 17-year-old Michelle Knapp, sold the asteroid and the car to collectors for $69,000. And in 1994 some fishermen saw another fireball, which exploded over

the western Pacific with the force of 100,000 tons of TNT, similar
to a nuclear detonation—so similar that there were reports that
U.S. national security officials woke up President Bill Clinton to
tell him about it.

And those are just a few examples of cosmic hits that were
actually witnessed; there were a lot more that weren't seen. Then
there were the asteroid hits that had been detected, but by people
who didn't want to talk about it—which is to say, the U.S. military. In 1993 the Defense Department finally released previously
classified data gathered over past decades by its Defense Support
Program—read "spy satellites"—and worldwide acoustic sensors
designed to detect missile launches and nuclear bomb tests. The
results were startling. The data indicated that a one-kiloton asteroid
airburst occurs somewhere over Earth every month or so, and a
15-kiloton burst—bigger than the Hiroshima bomb—happens on
average about once every year. Think about that. Almost every
year there's a Hiroshima-level explosion in the atmosphere that
under the right circumstances could devastate a city.

And that raised another potential danger from inbound Near-
Earth Objects—the possibility that one of them could spark a
nuclear war. Air Force Brigadier General Pete Worden of the U.S.
Space Command made the point after a small asteroid seared
through the atmosphere and exploded over the eastern Mediterranean with the force of an atomic bomb, creating a brilliant flash
and sending shock waves to the Earth's surface. At the time India
and Pakistan, both of them nuclear powers, were on the brink of
war, with their military forces on full alert. Although U.S. reconnaissance satellites quickly determined that the Mediterranean
explosion was a natural event, India and Pakistan did not have
that capability. In a 2002 speech Worden imagined what could
have happened if the bolide had exploded a few thousand miles
farther east.

"Imagine that the bright flash accompanied by a damaging shock wave had occurred over Delhi or Islamabad," Worden said. "The resulting panic in the nuclear-armed and hair-trigger militaries there could have been the spark that would have ignited the nuclear horror we've avoided for over half a century. That situation alone should be sufficient to get the world to take notice of the threat of asteroid impact."

But again, in the 1980s and early '90s hardly anybody seemed to care. The asteroid events that were witnessed generated some brief local or even national interest, but they were quickly forgotten. As with other disasters—wars, tsunamis, earthquakes—when it came to space-borne encounters, humans seemed to have short memories, and limited imaginations.

But then two things happened that finally concentrated the collective mind of humanity on the dangers looming in space.

First, somebody figured out what happened to the dinosaurs. And then something ran into Jupiter.

CHAPTER 5

T-REX WITH A
STRING OF PEARLS

In the mid-1970s a sandy-haired, bespectacled young Columbia University geologist named Walter Alvarez was poking around in the ancient rocks exposed in a gorge outside the small Italian town of Gubbio. He wasn't looking for anything having to do with asteroids; instead, he was interested in paleomagnetism, the study of rocks that show ancient changes in the Earth's magnetic field. But as he was peering at thick layers of limestone laid down over millions of years in the distant past, he noticed something unusual. There, sandwiched between the limestone layers, was a half-inch-thick layer of reddish brown clay.

That in itself didn't mean much. We've all seen how much of Earth's sub-surface is composed of many strata of different-colored sedimentary rock, like layers in a cake; one look at the Grand Canyon, or even a modest highway cut through a hill, tells you that much. But this was different. After he examined samples of the rocks more closely, Alvarez saw that the older limestone layers below the clay layer contained numerous fossils of long-extinct species of tiny sea creatures called foraminifera, or forams for short. But in the newer limestone layer directly above the thin layer of red clay there were hardly any of those foram species to be seen. And it was the same story with other rock samples Alvarez had taken in the area: lots of forams below the clay layer, hardly any above it. Clearly, something had happened around the time the clay layer was deposited that had caused those earlier species of sea creatures to go extinct.

And there was something else. It appeared that the thin layer of red clay had been deposited roughly sixty-five million years ago. That was at the end of what geologists call the Cretaceous Period, a time when dinosaurs roamed the Earth, and at the beginning of the Tertiary Period that featured the rise of the mammals—and, eventually, us. The so-called K-T Boundary between those geologic periods marked one of the greatest mass extinctions in Earth's history, an event that saw most of the world's plant and animal species wiped out, including almost all of the dinosaurs. (I say "almost all" because today's birds are descendants of species of avian dinosaurs.)

To Alvarez it raised an intriguing question: Could the thin red clay layer he found outside Gubbio somehow have been connected to both the extinction of the tiny sea creatures *and* the extinction of the mighty dinosaurs? And could that mass extinction have happened suddenly, out of the blue, over a relatively short period of time?

For a geologist—for almost any scientist of the day—it was a revolutionary thought. The prevailing wisdom was that the extinction of the dinosaurs and other species during the K-T extinction had taken at least a million years or so. There were various explanations for the dinosaurs' gradual demise: Climate change, declining sea levels, and a poisoned atmosphere from a long period of massive volcanic eruptions were all possible culprits. Other less conventional theories included chronic constipation caused by a change in dinosaur diet, and even a lowering of sperm counts in males caused by higher temperatures—what you might call the tighty-whities theory of reduced dinosaur reproduction. Still another suggestion was that after ruling the world for 170 million years or so the entire clade Dinosauria just gave up, victims of what's been called "paleoweltschmerz," a disenchantment with life in a difficult world. I imagine a lot of us humans occasionally feel the same way, especially after we watch the TV news.

The point is that virtually all of the then-prevailing serious theories concerning the K-T extinctions were based on the assumption that they had taken place gradually, over an extended period of time. And that assumption in turn was based on the scientific doctrine known as "uniformitarianism." That may sound like the dogma of a religious sect, but in fact it's a doctrine that dominated science throughout the nineteenth and twentieth centuries. Simply put, it held that all geological and biological changes are caused by known processes that occur over immense stretches of time. The rise and fall of the oceans, the buildup of ground by uplift and sedimentation, the evolution of species, the extinction of the dinosaurs and other ancient creatures—they were the result of slow, steady, even *dignified* natural processes. Sure, you might have your occasional violent hiccup, like volcanic eruptions or floods, but those were local events, and had no bearing on Earth's history.

At the opposite end of that uniformitarian doctrine was what was called "catastrophism," the notion that Earth's history had been periodically altered by sudden cataclysmic events that were beyond the current level of human experience—unprecedentedly gigantic floods, world-shattering earthquakes, enormous volcanic eruptions. It was and is a valid theory; we now know that Earth's geologic history has been punctuated by such catastrophic events. But unfortunately for scientific-minded catastrophists, by the mid-nineteenth century the theory of periodic worldwide cataclysms had been hijacked by religious forces who thought it fit nicely with their biblical views. Those fossilized sea shells found high up on mountaintops, miles away from the sea? To the religious catastrophists, they were put there by the Great Flood of Noah's time. Those recently discovered fossilized bones of strange creatures called "dinosaurs" that no longer existed on Earth? They were just animals that didn't make it onto the Ark. And so on. As a result of such theories, by the mid-nineteenth century most reputable scientists viewed catastrophism as a bad joke. It simply wasn't science.

Walter Alvarez understood the problem. As a geology student in the 1960s and early '70s he had been steeped in the uniformitarian view of slow, gradual change, and to challenge it by suggesting a sudden catastrophic extinction event would be—well, it would be scientific heresy. As Alvarez later recalled, "In the mid-1970s the thought of a catastrophic event in Earth history was disturbing. As a geology student I had learned that catastrophism is unscientific. . . . I had come to honor . . . the doctrine of uniformitarianism, and to avoid any mention of catastrophic events in Earth's past. But Nature seemed to be showing us something quite different."

But what kind of gigantic catastrophe could have caused the K-T extinction? What sort of cataclysm could be big enough and bad enough to wipe out 40-ton dinosaurs *and* microscopic sea creatures? At the same time? And around the world? Alvarez wasn't sure. So he decided to ask his dad for help.

Walter Alvarez's dad wasn't any ordinary dad. He was Luis W. Alvarez, a Berkeley physicist who was awarded a Nobel Prize in 1968 for his work in particle physics. During World War II Alvarez senior had also developed a series of new radar systems, worked on the Manhattan Project to build the first atomic bomb, and had even flown as a civilian observer on a B-29 accompanying the *Enola Gay* on its mission to Hiroshima. He also served on the CIA-sponsored Robertson Panel that investigated alien UFOs in

the early 1950s—the panel concluded they didn't exist—and later
he was called upon to examine scientific evidence in the John F.
Kennedy assassination. Most recently he had made an unsuccessful
attempt to use cosmic rays to detect hidden chambers in the Egyptian pyramid of Chephren at Giza. Like I said, no ordinary dad.

Over the next few years father and son kicked around various
extinction theories. At one point they thought the sudden K-T
extinction might have been caused by cosmic radiation from the
explosion of a distant supernova, but the science didn't work out
on that one. But finally it came to them. Based on the work of
Gene Shoemaker and others concerning Earth-impacting space
bodies, the Alvarezes realized that the only thing that could pack
enough punch to cause a sudden global extinction was an incoming
asteroid or comet. (Whether it was a comet nucleus or an asteroid
is still a matter of debate. There are good scientific arguments on
both sides, but for simplicity's sake I'll assume it was an asteroid.)

The scenario is now well known. Sixty-five million years ago
an asteroid six miles wide and traveling at least 40,000 miles per
hour blazed through the atmosphere and hit a shallow sea with
the energy equivalent of 100 million megatons of TNT—roughly
the equivalent of ten *billion* Hiroshima bombs—blasting out a
crater more than a hundred miles wide and several miles deep.
The resulting shock wave and fireball killed every living thing for
a thousand miles around, and sent a tsunami hundreds of feet
high sweeping over the entire region. The explosion sent chunks
of rock flying up into sub-orbital arcs, rocks that heated up like
returning Gemini space capsules as they plunged back through the
atmosphere, creating massive fireballs that started continent-wide
wildfires and filled the atmosphere with heavy black soot.

It got worse. Hundreds of cubic miles' worth of crushed rock
dust and vaporized asteroid rose into the stratosphere and spread
around the globe, mixing with the wildfire soot and blocking sun-

light, which disrupted food chains by preventing photosynthesis in plants on land and plankton in the seas. Temperatures initially dropped to freezing, followed by global warming as heat was trapped inside a shroud of dust and gases. Sulfuric acid in the dust soured the oceans and the land with acid rain. Among the dinosaurs that survived the initial blast effects, after the plants died the herbivores starved to death, then the carnivores that had preyed on the herbivores starved, too. Except for the aforementioned bird-like dinosaur species, every single dinosaur on the planet eventually was winked out.

Some species we still know today made it through the apocalypse. Alligators survived to eventually invade swimming pools in Florida, and sharks lived on to terrorize swimmers and moviegoers. Cockroaches, naturally, survived and prospered. On land some small, omnivorous, rat-like species of mammals cowered in their burrows, protected from the fire and the cold, and when they finally poked their little heads out of their holes there was plenty of rotting dinosaur meat to dine on until things settled down. For the previous hundred million years of their existence these small creatures had had to worry about getting eaten or stepped on by dinosaurs. With that threat eliminated, now it was their time to shine—which was a good thing for us, since those small mammals were in a sense our great-great-great-to-the-nth-power-grandparents. If the dinosaurs hadn't being wiped out, we humans almost certainly wouldn't be here.

But survival was the exception, not the rule; three-quarters of all the species on Earth disappeared.

(If a similar-sized object were to hit Earth today it could wipe out a continent, killing hundreds of millions of people in the initial blast. The aftereffects on the climate would be global, with agriculture disrupted for years and billions of people threatened with starvation. It would end civilization as we know it, as least for a long time. But it probably wouldn't cause humans to go extinct. Some people outside the immediate impact zone almost certainly would have the knowledge and technology—and enough stored food supplies—to carry on. It would be the worst day in the history of the human species, but chances are it wouldn't be the last.)

The Alvarezes weren't the first scientists to consider the destructive potential of a massive cosmic impact on Earth. Way back in the eighteenth century Edmond Halley had speculated that a comet impact might have caused Noah's flood by making the oceans slosh over the globe. In 1953 two guys named Allan O. Kelly and Frank Dachille wrote a book called *Target: Earth* that suggested the dinosaurs had been wiped out by a meteoric impact, but perhaps because the lead author was a California rancher and self-taught geologist, and the book was self-published, no one paid attention. More recently, in 1969 a respected paleontologist named Digby McLaren had speculated at a paleontologists' conference that an earlier mass extinction 250 million years ago had been caused by a giant meteoritic impact—although most of the audience members apparently thought he was only joking. There were a few others as well.

But what set the Alvarez hypothesis apart was that it wasn't just a theory. They actually had *evidence* of a massive meteoric impact that coincided in time with the K-T extinction—cold, hard, scientifically testable evidence. It came in the form of iridium—that same valuable meteoric metal that had so excited Daniel Barringer decades before.

You'll recall that iridium is a hard, silvery metal that's extremely rare on Earth's surface but is relatively abundant in asteroids and meteorites. In an effort to determine how long it had taken for the red K-T Boundary clay to be deposited, the Alvarezes submitted a sample of the clay to UC Berkeley chemists Frank Asaro and Helen Vaughn Michel to check for its iridium content. And what they found was astonishing. The iridium levels in the clay were much higher than they normally should have been—a hundred times or more higher, as it turned out. And it was the same thing with other K-T Boundary clay samples collected from other parts of the world; they were chock-full of iridium as well.

The conclusion was inescapable: When the iridium-bearing asteroid hit the Earth and vaporized itself sixty-five million years ago it sent up an enormous cloud of relatively iridium-rich dust into the atmosphere, dust that eventually settled in thin clay layers around the globe, like the one Walter Alvarez found in Italy and others had found elsewhere. No terrestrial geologic force could have deposited that much iridium in so short a time; it had to have been put there by a massive asteroid impact. And the global effects of that massive impact had caused the K-T extinction, almost overnight.

The Alvarez team formally announced their impact-extinction theory in the June 1980 issue of the journal *Science*. Titled "Extraterrestrial Cause for the Cretaceous-Tertiary Extinction," the 14-page article didn't discuss tsunamis or wildfires or acid rain; all that would come later. Instead, the Alvarezes simply postulated that the impact of a six-mile-wide asteroid had enveloped the Earth in a sun-blocking dust cloud that caused the mass extinction. The paper barely mentioned the demise of the dinosaurs, concentrating instead on the sudden extinction of much more populous microscopic sea creatures—the forams—as evidence of their theory. But the brief allusion to dinosaurs was enough.

A hurtling space rock wiped out the dinosaurs! It was a sensation.

It's hard to imagine a more perfect popular-science news story. After all, there was hardly a kid or former kid in America who hadn't had at least a passing interest in dinosaurs. Dinosaurs were everywhere—in toys, books, movies, TV shows. A guy in a dinosaur suit smashing a scale-model Tokyo in *Godzilla*. The fabulously pulchritudinous—and barely dressed—Raquel Welch battling dinosaurs in the 1966 film *One Million Years B.C.* Fred Flintstone operating a brontosaurus crane at the Slate Rock and Gravel Company and being greeted by dog-like Dino the pet dinosaur when he got home. You couldn't buy a gallon of high-octane at the Sinclair station without seeing a green *Apatosaurus* on the company logo. And of course there were all those school field trips to the natural history museums with their full-sized reconstructions of the dreaded T-Rex.

At the same time, there was hardly a kid or former kid who hadn't also had a passing interest in asteroids. As I mentioned earlier, asteroids were part of the culture, from *Star Wars* to Atari video games. And when you could combine asteroids with dinosaurs and a world-shattering violent catastrophe, well, it was irresistible. The popular press quickly latched on to the Alvarez theory, with newspapers bearing headlines like "The Day the Dinosaurs Died" and magazine cover stories depicting dinosaurs facing their imminent doom. A *Time* magazine cover story on the Alvarez theory portrayed a somewhat portly green-and-yellow T-Rex peering nervously over his shoulder at a fiery mushroom cloud, while other renditions showed alamosauruses and triceratopses futilely fleeing from an enormous approaching fireball. The public couldn't get enough of this stuff.

The reaction to the Alvarez team's theory among the scientific establishment was dramatic as well—dramatically negative. Of course, the Alvarezes never expected that hordes of excited paleontologists and geologists would stream onto the Berkeley campus, hoist Luis Alvarez et al. up on their shoulders and then triumphantly parade them through the streets amid joyous cries of "Thank you! Thank you!" and "Eureka! You have found it!" Science doesn't work that way, especially when confronted with a groundbreaking new theory. But even the Alvarezes were stunned by the level of controversy their theory generated.

True, some scientists were receptive to the impact-extinction theory; certainly Gene Shoemaker and other impact researchers were well aware of the destructive power of an asteroid strike. But many, perhaps most people in the science world were initially skeptical, if not outright hostile. That was particularly true among paleontologists—which perhaps was understandable. After all, they had spent entire careers studying the K-T extinction based on good, solid uniformitarian principles, painstakingly examining the fossil records of *Globigerina eugubina* and *Guembelitria cretacea* and other obscure creatures and publishing the results in scholarly journals that no one except a handful of other paleontologists ever read. And for decades the paleontology community had been confidently asserting that the dinosaurs and other Cretaceous species had slowly gone extinct over hundreds of thousands if not millions of years. But now this Alvarez bunch, led by a physicist who probably didn't know a foram from a dryptosaurus, suddenly comes along with this wild new catastrophist theory that was making them look like idiots. And the press and public were eating it up! Those damned Alvarezes were rock stars! They and their ridiculous theory were on the cover of *Time* magazine, for God's sake! It was infuriating.

(Not all of the news media signed on to the Alvarez theory. For example, the staid *New York Times* covered the impact-extinction debate objectively in its news pages, but its editorial page took the cautious uniformitarian view—with a pointed dig at proponents of the impact theory. In an unsigned editorial in 1985, titled "Miscasting the Dinosaur's Horoscope," the *Times* said, "Terrestrial events, like volcanic activity or changes in climate or sea level, are the most immediate possible causes of mass extinctions. Astronomers should leave to astrologers the task of seeking the cause of earthly events in the stars." *Astrologers!* To people who studied cosmic impacts and other space sciences, seeing the words "astrologers" and "astronomers" in the same sentence in the august *New York Times* was a mortal insult.)

Initially the scientific community's public pushback to the Alvarez theory came in properly decorous forms. Scholarly papers questioned the Alvarez team's facts and conclusions, arguing that other evidence convincingly showed that the K-T extinction had indeed been a slow, gradual process. As one scientist gently put it, "It does not seem *necessary* to invoke an unusual event to account for the demise of the dinosaurs" (emphasis added).

But as the Alvarezes and their theory continued to enjoy a heady dose of popular acclaim, things soon turned ugly—and personal. As one noted dinosaur authority told the *New York Times*, "The arrogance of those people is simply unbelievable. . . . In effect, they're saying, 'We high-tech people have all the answers, and you paleontologists are just primitive rock hounds.'" Another respected paleontologist publicly dismissed the Alvarez theory as "codswallop," meaning nonsense—and that was one of the softer characterizations. Eventually it got to the point where a prominent Dartmouth College scientist publicly excoriated the Alvarez theory as "degenerative," "pathological," "dangerous," and suggested that it was "some kind of scam." There were even dark rumors that Luis Alvarez was trying to scuttle the professional prospects of those who disagreed with him; one professor claimed that Alvarez senior had confronted him at a conference and essentially issued a Mob-style warning, something along the lines of, "Nice little career you've got going here, pal. Be a shame if anything happened to it."

The affable Walter Alvarez generally tried to play down the personal attacks. But his father was a different kind of Alvarez. Everyone agrees that the senior Alvarez had a healthy ego; his detractors, who were many, would argue that his ego was positively bursting with health. He could be witheringly caustic about those whom he thought were his intellectual inferiors—which is to say, just about everybody. His status as a physicist may have contributed to that attitude. Every profession has its pecking order, and at the top of the scientific pyramid in the twentieth century were the physicists, at least in their own minds, with every other scientific discipline a cut or more below. In fact, there's a famous story about 1920s Austrian physicist Wolfgang Pauli that illustrates the point. When his cabaret-dancer wife left him for another scientist, Pauli supposedly told a friend, "Had she taken a bullfighter I would have understood. But a *chemist?*"

In his disdain for other, seemingly lesser scientists, Professor Pauli had nothing on Luis Alvarez. And in a 1988 interview with the *New York Times* about his scientific opponents, Alvarez let his scorn and disdain out for a romp.

"I don't like to say bad things about paleontologists," Alvarez told the *Times*—and then he went on to do just that, adding, "but they're really not very good scientists. They're more like stamp collectors." He also characterized a specific opponent as a "weak sister," another as a laughingstock within the scientific community, and then concluded with "I don't want to hold up these guys to too much scorn. But they deserve some scorn, because they're publishing scientific nonsense." As for rumors he had tried to damage the careers of his opponents, Alvarez denied it—although not very convincingly.

(When he gave his blistering interview Luis Alvarez was suffering from terminal esophageal cancer. As he put it, "I can say these things . . . because this is my last hurrah, and I have to tell the truth." He died that same year.)

All in all, the dispute over the Alvarez extinction theory was probably the most publicly rancorous scientific debate since Darwin. But despite all the unseemly attacks and insults from both sides, there was science being accomplished. In the 1980s alone more than *two thousand* scientific papers were published on some aspect of the Alvarez impact-extinction theory, many of them bearing such eye-glazing titles as "Gradual Dinosaur Extinction and Simultaneous Ungulate Radiation in the Hell Creek Formation," and "Paleoenvironments of the Vertebrate-Bearing Strata During the Cretaceous-Paleocene Transition." Some of these papers advanced the Alvarez theory with new scientific test results, while others advanced the theory simply by failing to disprove it.

Meanwhile, other evidence of a giant cosmic impact sixty-five million years ago was piling up. Dozens more iridium-rich K-T Boundary clay deposits were discovered around the world. Traces of meteoric impact materials that matched the time frame started showing up at various locations—shocked quartz, heat-formed glass spherules called tektites, our old pal coesite, and so on. Researchers also found evidence that a giant tsunami had indeed swept over the Gulf of Mexico region at the time of the K-T extinctions.

Still, there was one big hole in the Alvarez theory. That hole was, literally, a hole—or rather the lack of one. If in fact a six-mile-wide asteroid had caused the K-T extinction, where was the crater? Where was the 100-mile-plus-wide, 65-million-year-old crater such an impact would have blasted out of the Earth? None of the then-known impact craters seemed to fit the bill in either size or age. So where was it? The Alvarezes couldn't say.

The fact that they didn't have a crater didn't disprove the impact-extinction theory. The K-T crater could have been part of the 20 percent or so of the ocean floor that over 65 million years had been destroyed by geologic sea floor spreading and subduction. Maybe the crater was buried under mile-thick ice in Antarctica, or covered up by sedimentation in some remote, unmapped portion of the ocean floor, or maybe it had been filled in and scraped flat by glaciation. There were plenty of reasons why it hadn't been found, or why it might no longer exist. But the lack of the so-called smoking gun gave Alvarez opponents another line of attack.

And then, through no small amount of luck, researchers found the crater. In 1978 an American geophysicist named Glen Penfield, then employed by the Pemex Mexican oil corporation, was looking at old magnetic and gravity surveys of the sea floor near the Yucatán Peninsula when he noticed a large, circular geologic structure buried thousands of feet underground and underwater near the Mexican coastal village of Chicxulub. He and a colleague, Antonio Camargo, later presented a paper suggesting that it could be an impact crater, but in those days before Google the paper somehow escaped the notice of the Alvarezes and other researchers. It wasn't until a decade later, after Penfield linked up with K-T crater researchers Alan Hildebrand and David Kring (among others), that they finally made the case. Based on old drill core samples and meteoric materials found in abundance near the Chicxulub crater, and the crater's size and age, the researchers concluded that it fit perfectly with the Alvarez K-T extinction theory. In 1991 they announced that the Chicxulub crater was indeed the smoking gun that had put a bullet into the dinosaurs.

And as it turned out, the asteroid that made the Chicxulub crater couldn't have picked a worse spot to hit, at least as far as the dinosaurs and other Cretaceous creatures were concerned. If it had hit in the deep ocean it would have created an enormous fireball and sparked a mega-tsunami of biblical proportions, but it probably wouldn't have put as much dangerous dust into the atmosphere. For dinosaurs and other species it wouldn't have been a day at the beach, but it might not have wiped them out. But instead the asteroid hit a portion of the North American continental shelf that was rich in carbon-dioxide-bearing limestone and sulfur-bearing gypsum, producing especially lethal clouds of dust. Again, bad for the dinosaurs, good for us.

As you might expect, there were still scientists who weren't on board, Chicxulub or no Chicxulub. To this day there is continuing debate among some scientists as to whether the Chicxulub impact or any other cosmic impact was a significant factor in the K-T extinction. A few argue that Chicxulub isn't an impact crater at all. Others say it is an impact crater but that the asteroid hit Earth three hundred thousand years before the K-T extinction. Still others say that the deadly effects of the impact were merely the final straw for dinosaurs and other diminishing species that had been subjected to hundreds of thousands of years of climate change caused by enormous volcanic eruptions in what is now India. And on and on.

As I said in the introduction, it's not my purpose here to litigate the controversy. But with all due respect to the skeptics, it's probably fair to say that most scientists today accept the impact of an asteroid or comet as the predominant cause of the K-T extinction.

And yet, even as the impact theory drew increasing support in the 1980s and early 1990s, it was still hard for a lot of scientists to get their minds around the concept. Sure, they understood the mechanics of hypervelocity impacts. They could crunch the numbers on the enormous energy released by the Chicxulub impact, and mathematically calculate the catastrophic blast and atmospheric effects thereof. It all made scientific sense—but somehow they just couldn't quite *feel* it. Emotionally it was hard to grasp how such a relatively small impactor could deliver such a titanic Earth-changing blow. After all, the Earth is eight thousand miles in diameter, and the Chicxulub asteroid was only about six miles in diameter; by volume, the Earth is more than *two billion times* bigger than the asteroid that ran into it. Proportionally it was like a piece of pea gravel hitting a 100-foot-wide boulder. And of course, no one in recorded human history had ever seen such an event, on Earth or anywhere else; it was hard to even imagine it.

But then, in a coincidence that almost strains credulity, the cosmos arranged a demonstration. Next to an actual asteroid impact on Earth, it was about as dramatic a demonstration of the power of cosmic collisions as there could be. And as with so much else in the history of space impacts, Gene Shoemaker was in on it.

O

In the first decade of its existence Gene and Carolyn Shoemaker's Palomar Asteroid and Comet Survey, operating from the Palomar Observatory outside San Diego, had discovered several thousand new asteroids, including more than forty potentially hazardous Earth-crossers, and more than two dozen new comets. In fact, at the time, Carolyn Shoemaker had discovered more new comets than any woman in history—and eventually she discovered more comets than anyone else in history up to that time, male or female.

By the standards of the day the Shoemakers were making a lot of discoveries, but it remained slow and laborious work—cutting and loading film into the 18-inch telescope, aiming the telescope by hand, taking multiple exposures of a patch of sky, developing the film, and then staring at it through a stereomicroscope for telltale images of asteroids that seemed to "float" above the starry background. The survey also remained a poorly funded, shoestring operation, with the Shoemakers being assisted by a number of generally unpaid volunteers—among them a 44-year-old Canadian named David Levy, a college English Lit. major turned serious amateur astronomer who had already discovered several comets on his own.

As Levy later recalled, on the afternoon of March 25, 1993, Carolyn Shoemaker was sitting at a desk at the Palomar Observatory, staring through a stereomicroscope at some photographic images of a patch of sky taken with the 18-inch telescope the night before. It had been a tough observation run, plagued by cloudy weather, accidentally ruined film, and other small catastrophes. Almost a year had gone by since their last comet discovery, which was discouraging; it had gotten to the point where Carolyn ruefully complained that "I *used* to be a person who found comets." No one expected too much in the way of results from the previous night's exposures.

But then Carolyn saw something on one of the films. Suddenly she sat up straight in her chair and told her husband and Levy, "There's something strange out there . . . it looks like a squashed comet."

A *squashed* comet? There was no such thing. It had to be a distant galaxy, or some dust or defect on the film. But when Levy and Gene Shoemaker looked at the image it clearly showed a long, fuzzy bar of light—that was the squashed part—with unmistakable comet tails sticking out of it. Further examination showed that the bar of light was in fact a series of glowing comet nuclei arranged in a line—what came to be known as the "string of pearls." Because this was the ninth periodic comet discovered by the Shoemaker-Levy team it was officially named Comet Shoemaker-Levy 9—and in decades of searching for asteroids and comets, it was the most unusual thing any of them had ever seen in the sky. As Levy later put it, "We hadn't just found a comet; we'd found a real unicorn!"

And the story of Shoemaker-Levy 9 soon got stranger still. Subsequent observations by astronomers around the world confirmed that the small comet fragments that formed the string of pearls weren't orbiting around the Sun. Instead, they were orbiting around *Jupiter*—and in a little over a year they were going to collide with a planet. The good news was that for the first time in recorded history, scientists would actually be able to see a cosmic collision. And the really good news was that it was massive Jupiter, not Earth, that was going to take the hit.

Next to the Sun, Jupiter is the big boy of the Solar System. It's some 88,000 miles in diameter, eleven times bigger than Earth; in terms of mass Jupiter is more than twice as big as all the other planets combined. Composed primarily of hydrogen and helium that's wrapped around a small, solid core, Jupiter's outer atmosphere displays a series of multicolored horizontal jet streams of gases whose edges violently swirl around like hurricanes. It's beautiful, but with an average temperature of minus 234 degrees Fahrenheit and heavy concentrations of ammonia in the atmosphere you certainly wouldn't want to live there.

Because of its massive size and correspondingly enormous gravitational pull, Jupiter plays a key role in Solar System mechanics—and from Earth's perspective, it's a mixed bag. On the one hand, ever since the Solar System formed more than four billion years ago, Jupiter's gravity has been sucking in countless asteroids and comets that otherwise could eventually have collided with Earth; it has also been flinging some potentially Earth-bound comets out of the Solar System altogether. Without Jupiter there'd be a lot more dangerous asteroids out there than there are today, and we might be getting hit by a world-buster every few centuries instead of every million years or so. On the other hand, Jupiter's gravity also helps disturb the orbits of some incoming comets and some asteroids in the main asteroid belt, sending them hurtling into Earth's neighborhood—with potentially catastrophic results. On the *other* other hand, if it weren't for Jupiter's gravitational forces, all of the asteroids in the main belt might have coalesced into a single planet long ago, which means we wouldn't have to worry about asteroids at all. The bottom line is that in terms of Earth-impacting asteroids and comets, Jupiter is sort of our friend—but it also has a lot to answer for.

As for Comet Shoemaker-Levy 9's relationship with Jupiter, apparently the previously Sun-orbiting comet had been "captured" by Jupiter's gravity sometime in the mid-1960s and it had been orbiting around Jupiter ever since then without anyone on Earth spotting it. Originally the comet nucleus was about three or four miles wide—half the size of the Chicxulub asteroid or comet. But during one particularly close pass in 1992, Jupiter's enormous gravity caused the comet's loosely packed nucleus to stretch out like a noodle and then break into almost two dozen pieces, the biggest of which was about a mile in diameter; the pieces were officially dubbed Fragments A through W. The best orbital calculations showed that over a weeklong period in July 1994 those fragments, traveling at a blistering 134,000 miles per hour, would slam into Jupiter's atmosphere one after another, like bullets from a machine gun.

For Gene Shoemaker, it was an affirmation of his life's work on cosmic collisions—and so stunning that he could hardly grasp it. "I don't believe it," he said when the predicted chances of a collision went from possible to certain. "In my lifetime we're going to see an impact." It was, he said, "a bloody miracle."

But what the effect of the impacts would be was anybody's guess. Remember, comets are loosely packed dust and ice, and they can easily disintegrate; it's been jokingly suggested that you could blow one apart with a sneeze. Remember also that all of the comets in the string were much smaller than the Chicxulub impactor, and they were hitting the second-biggest thing in the Solar System. And they wouldn't be hitting a solid object like Earth, but instead a Jovian atmosphere some three thousand miles thick. Given all that, more than a few scientists thought the fragments would disappear into Jupiter's atmosphere with barely a ripple, like rocks thrown into a fluffy snowbank; at the most they might produce some small meteor showers in Jupiter's atmosphere.

Shortly before the impacts the science journal *Nature* published an article about the impending event titled "The Big Fizzle Is Coming"—and even the Shoemakers and Levy were worried that the headline might be right.

But the popular press had no such reservations. Like the Alvarez dinosaur extinction story a decade earlier, the impending cosmic collision was too good a story to muck it up with a lot of scientific "maybe this" and "maybe that"; headline writers demanded that it be spectacular. For example, a *Time* magazine cover story illustration showed a string of comets closing in on Jupiter under the headline "Cosmic Crash," accompanied by the notation, "A shattered comet is about to hit Jupiter, creating the biggest explosion ever witnessed in the Solar System." A *New York Times* headline declared, "Comet to Hit Jupiter with Texas-Sized Bang; Astronomers Gear Up."

And gear up the astronomers did. Virtually every professional and amateur astronomer on Earth was getting ready to watch the impacts, whether from a giant telescope in a mountaintop observatory or a small hobby telescope in a backyard. In space, the Hubble Space Telescope in low orbit around the Earth was trained on Jupiter, as was the German Röntgensatellite orbiting X-ray telescope. The *Galileo* unmanned spacecraft, which had been launched for a Jupiter probe even before Shoemaker-Levy 9 was discovered, was also prepared to take a gander at the Jupiter impacts while it was still 150 million miles away. The joint NASA/European Space Agency Sun-orbiting spacecraft *Ulysses* was in on the game as well, and even the deep-space *Voyager 2* spacecraft was set up to monitor radio emissions from the impacts from more than four billion miles away. Jupiter was thoroughly covered.

And when the impacts began on July 16, 1994, a fizzle they were not. "Fragment A," one of the smaller fragments, hit Jupiter's atmosphere at 37 miles per second, sending a 40,000-degree fireball plume hundreds of miles above Jupiter's cloud layer and leaving a scar on Jupiter's atmosphere that was visible with Earth telescopes when the planet's rotation brought it into view. The largest impactor, "Fragment G," was even more astonishing. It hit Jupiter with the energy equivalent of *six million megatons* of TNT, some six hundred times the energy of all the nuclear weapons on Earth at the height of the Cold War. "G" also scorched out a scar some 7,000 miles wide that was easily visible from backyard telescopes. Other impacts left similar scars visible for months.

The telescopic images of the impact event were jaw-dropping, for scientists and the public as well. Every TV network gave the impacts major play, and newspapers erupted with front-page headlines like "Comet Crashes into Jupiter in Dazzling Galactic Show" and "Comet Scars Jupiter with Earth-sized Blot."

(As for the initial skeptics, at least some of them were good-natured about having been so wrong. Scientist Paul Weissman, author of the "Big Fizzle Is Coming" article in *Nature*, cheerfully noted that "Everyone knows that 'fizzle' is a Yiddish word meaning 'Great big humongous Jupiter-shaking comet explosion.'")

For the world in general the Jupiter impacts were a kind of "Holy shit!" moment. And it inevitably raised the question: If it happened to Jupiter, could it happen to us?

Actually there were a few people, a very few, who for years had been warning that yes, it certainly could happen to us. There was the Shoemaker-Levy team, of course, as well as planetary scientists and astronomers like Clark R. Chapman, David Morrison, Glo Helin, and others. Prompted by the Alvarez extinction theory, through the 1980s and early '90s there had been a series of scientific conferences and workshops on the asteroid threat, with names like "Collision of Asteroids and Comets with the Earth: Physical and Human Consequences" and "Workshop on Hazards Due to Comets and Asteroids." The attending scientists had concluded that there was a "small but non-zero" risk of a civilization-threatening impact on Earth in the unforeseeable future—"non-zero" meaning the risk was clearly there but impossible to precisely quantify. They recommended that a worldwide network of telescopic sky surveys be set up to expand the search for potentially hazardous Near-Earth Objects—a network eventually known as "Spaceguard." But the suggestion didn't gain much traction.

There had also been a few more people getting into the active search for Near-Earth Objects. One of the most notable was the aforementioned Tom Gehrels, a teenaged World War II Dutch resistance fighter and British commando turned astronomer. After earlier work on asteroids in the 1950s and '60s, in 1980 Gehrels and astronomer Robert S. McMillan launched what they called the "Spacewatch" sky survey, operating from the Kitt Peak National Observatory on top of a 7,000-foot mountain southwest of Tucson. Like the Shoemakers' Palomar sky survey, Spacewatch's mission was to find potentially hazardous asteroids. But what set Spacewatch apart was new technology—the same technology that eventually killed film cameras and filled up our smartphones with selfies and countless pictures of our grandkids. For the first time, Spacewatch used telescopes equipped with "charge-coupled devices"—CCDs—to electronically produce images of Near-Earth asteroids and other space bodies. No longer would astronomers have to deal with bulky and slow photographic film, and in the process maybe, just maybe, discover a new Near-Earth asteroid every few months, or even more rarely a new comet. With CCDs and computers linked to their telescopes, astronomers would capture digital images of first dozens, then hundreds, then thousands of asteroids in a single run of observing. It revolutionized the asteroid-searching business.

Still, during the 1980s and early '90s the search for hazardous NEOs remained confined to a vanishingly small number of people.

One scientist famously observed that there were more people working a daytime shift at a single McDonald's than there were people actively engaged in searching for potentially hazardous asteroids—an exaggeration maybe, but not by much.

Part of the problem was lack of funding by government agencies like NASA. In a 1991 funding authorization for NASA a congressional committee declared: "The chances of Earth being struck by a large asteroid are extremely small, but since the consequences of such a collision are extremely large, the Committee believes that it is only prudent to assess the nature of the threat." The committee urged NASA to develop "a program for dramatically increasing the detection rate of Earth-orbit-crossing asteroids." But while the space agency was happy to sponsor scientific conferences and reports on the asteroid threat, it was extremely reluctant to fork over any serious money to actually investigate the threat. For example, in 1993, the year they discovered Shoemaker-Levy 9, the Shoemakers' sky survey received zero funding from NASA, and Tom Gehrels's Spacewatch survey was almost completely paid for with private grants and donations, with only small grants from NASA.

And there was another problem. The search for Near-Earth Objects was also hard up against what was known as the "giggle factor"—the eye-rolling, you-gotta-be-kidding-me reaction to the notion that any serious person would spend his or her time looking for giant rocks plummeting down from space. Sure, people could embrace the concept of a monster asteroid wiping out the dinosaurs—but that was sixty-five million years ago, for goodness' sake; things like that couldn't happen now. To a lot of people, the whole idea of searching for catastrophic cosmic impactors seemed—well, it seemed silly, a waste of time and money, like running around shouting, "The sky is falling! The sky is falling!" It was Chicken Little stuff.

As one contemporary research paper put it, "Clearly, elected officials in Washington are not being inundated with mail from constituents complaining that a member of their family has just been killed or their property destroyed by a marauding asteroid. . . . The prevailing view among government officials who hear about this issue for the first time is that the epoch of large asteroid strikes on Earth ended millions or billions of years ago."

And it wasn't just the general public and government officials who thought that way. Much of the scientific community thought so, too.

"The potential of an asteroid hitting the Earth and destroying civilization was viewed as far out, comic-book-fantasy-science-fiction-type thinking," recalls Bob McMillan, co-founder of the Spacewatch survey. "The asteroid hazard simply wasn't considered to be a scientific priority. That was a huge impediment to getting significant funding."

The Shoemaker-Levy 9 impacts on Jupiter in 1994 helped change that. They demonstrated for all the world to see that cosmic impacts of world-shattering proportions aren't relics of the ancient past, something from the time of the dinosaurs; they could happen now, in real time. Almost everybody now understood that but for the grace of God or gravity, that string of giant space pearls that blasted Jupiter could just as easily have blasted civilization off the face of the Earth.

"The solar system no longer seems quite so far away as it did before," planetary scientist Kevin Zahnle of NASA's Ames Research Center said after the impacts. "Here we are, close to the edge, protected from the true enormity of the universe by a thin blue line. A day will come when the sheltering sky is torn apart with a power that beggars the imagination. It has happened before. Ask any dinosaur, if you can find one. This is a dangerous place."

As for the giggle factor, after the Jupiter impacts the growing consensus was that maybe it was time to stop giggling and start getting serious about this cosmic-impact thing—an attitude summed up by a *New York Times* headline: "When Worlds Collide: A Threat to Earth Is a Joke No Longer." "The Chicken Little crowd, which once drew smiles by suggesting that Earth could be devastated by killer rocks from outer space, is suddenly finding its warnings and agenda taken seriously now that Jupiter has taken a beating in recorded history's biggest show of cosmic violence," the accompanying article said.

David Levy saw the change in attitudes as well, noting that "The giggle factor disappeared after Shoemaker-Levy 9." He was right in the sense that the scientific community started taking the impact threat more seriously. But as we'll see, the giggle factor concerning cosmic impacts unfortunately remains alive in some quarters to this day.

As for Gene Shoemaker, he can be forgiven for a little bit of post-impact I-told-you-so crowing.

"Yes, Virginia, comets really do impact planets, and the Earth really is at risk," he said in an interview. "I think the fact that we really have now seen an impact, in fact a whole series of impacts, on Jupiter this week has changed a lot of minds."

Politically, the Jupiter impacts marked a turning point. While the string of impacts was under way, a U.S. House of Representatives committee on science directed NASA—not urged but directed—to address the asteroid and comet threat; NASA responded by forming a committee to study the problem, headed by Gene Shoemaker. After several more years of scientific conferences and workshops and studies, in 1998 Congress formally tasked NASA with implementing the long-sought "Spaceguard goal" of finding and tracking 90 percent of the potentially catastrophic Near-Earth Objects within the next decade—"catastrophic" NEOs being defined as those a half-mile wide or larger that could cause worldwide destruction or critical environmental damage. Later the search would be expanded to Near-Earth Objects bigger than about 150 yards across, objects that could cause massive local or regional destruction, with the goal being a 90 percent detection rate by 2020. The idea was to put together a database of all the asteroids that might one day threaten Earth, sort of like explorers in the eighteenth and nineteenth centuries charting the reefs and shoals in the world's oceans. Once we know where those asteroidal reefs and shoals are, we can avoid them—or rather, as we'll see in a later chapter, we can figure out a way to make them avoid us.

It may sound silly, but it didn't hurt that 1998 also saw the release of two summer blockbuster movies based on cosmic impacts—*Armageddon* and *Deep Impact*—that helped focus public attention on the issue. As you probably know, *Deep Impact* featured Robert Duvall using a nuclear weapon to blow up a world-destroying comet, with partial success; a piece of it still hits the ocean and the resulting tsunami destroys the East Coast, but humanity at large is saved. It grossed $350 million worldwide. *Armageddon* featured Bruce Willis leading a group of misfit oil-well drillers into space to blow up an asteroid "the size of Texas" that was heading toward Earth. It grossed a half billion dollars at the box office, but scientifically it was execrable. In fact, there were rumors, perhaps of the urban legend variety, that NASA used *Armageddon* as a training device, showing it to management trainees to see if they could spot the reported 168 scientific impossibilities in the movie.

Anyway, the Jupiter impacts changed the way the world looked at cosmic collisions. A quarter century after Gene Shoemaker launched the first tentative attempts to find and track Earth-threatening asteroids and comets, a new era of asteroid detection and exploration was beginning.

But sadly, Shoemaker wouldn't be there to see it.

On July 18, 1997, Gene and Carolyn Shoemaker were on one of their frequent expeditions to the Australian Outback, an expanse of empty, mostly desert-like country that is rich in both known and perhaps undiscovered impact craters. It was the sort of place where you could drive all day without seeing another vehicle. And yet, as they were on their way to the ancient Goat Paddock impact crater, driving an old Toyota Hilux four-wheel-drive truck on a washboard dirt track outside of Alice Springs, the Shoemakers crested a low rise and their truck collided with an oncoming Toyota Land Cruiser. Gene Shoemaker was killed instantly. Carolyn was seriously injured but eventually recovered. (The occupants of the Land Cruiser were not seriously injured.)

No one missed the irony. After devoting most of his professional life to cosmic impacts, Gene Shoemaker died because two small moving objects in a vast empty space had somehow managed to run into each other.

Shoemaker's sudden death at age sixty-nine made headlines around the world, and sent much of the scientific community into shock. I've spoken with people who knew him who still grow quiet when they remember it. Over the course of his long career Shoemaker had revolutionized the study of cosmic collisions, won more than thirty prestigious scientific awards, had been publicly feted by American presidents and other heads of state and had influenced a generation of young scientists. No one could believe that he was really gone.

Shoemaker was cremated, and most of his ashes were scattered at the place he loved best on Earth—Meteor Crater. But it's a measure of Shoemaker's standing in the world of planetary science that in 1998 NASA enthusiastically agreed to have a small container of his ashes placed aboard the *Lunar Prospector* unmanned Moon-probe spacecraft. A laser-etched brass foil wrapping around the container displayed an image of a comet, a picture of Meteor Crater and a quote from *Romeo and Juliet*. After completing its mission in 1999 the spacecraft was intentionally crashed into an impact crater near the lunar south pole, carrying Shoemaker's ashes with it. It was the first and so far the only time that a human being had ever been interred on another celestial body.

It was a nice ending. Gene Shoemaker had finally made it to the Moon.

○

In the years after Shoemaker's death we finally started getting serious about asteroids. A series of unmanned probes were rocketed into space to take a close-up look at these mysterious objects, and new asteroid search programs sprung up to try to get a handle on how many of these things are out there—and to figure out if any are headed our way.

As I mentioned earlier, the *Galileo* unmanned spacecraft had already taken the first-ever photographs of asteroids when it zoomed by asteroids Gaspra and Ida on its way to orbit Jupiter in the early 1990s. The 5,600-pound space vehicle, which looked like something made with an erector set and an umbrella, passed less than a thousand miles from Gaspra, a rocky asteroid shaped like a shark's tooth, and sent back dozens of pictures of its heavily impact-scarred surface. *Galileo* went on to pass within 1,500 miles of asteroid Ida, in the process discovering that the 40-mile-wide asteroid has its own satellite, a mile-wide little "moon" called Dactyl, which orbits around Ida every twenty hours or so. It was the first confirmed discovery of a "binary" asteroid system, and since then more than three hundred asteroid "moons" have been discovered—although there are certainly many, many more out there.

Asteroids Gaspra and Ida are both in the far-off main asteroid belt, but through a series of unmanned missions we also got our first look at asteroids closer to home—the Near-Earth Objects that someday could pose a threat to Earth. In 1996 NASA launched the Near-Earth Asteroid Rendezvous (NEAR) unmanned spacecraft mission—subsequently renamed the NEAR Shoemaker mission in Gene Shoemaker's honor—to take a close-up look at Near-Earth asteroid Eros, a ten-mile-wide space rock that's shaped like a lady's shoe. NEAR Shoemaker spent a year orbiting Eros and then in 2001 the spacecraft actually "soft-landed" on its surface. It was the first time ever that a man-made object had orbited or landed on an asteroid. Despite the harsh surface conditions—temperatures on the rotating asteroid range from 200 degrees Fahrenheit in sunlight to more than 200 degrees below zero in the dark—the spacecraft continued to send back signals for almost two weeks before finally shutting down. The NEAR Shoemaker vehicle is still up there, riding Eros through space, and presumably it will be parked there forever. And for now the parking is free. Shortly after NEAR Shoemaker landed, a Nevada-based "space activist" named Greg Nemitz announced that he had filed an ownership claim on asteroid Eros and sent NASA a bill for $20 in "parking and storage fees" for the inert spacecraft. NASA refused to pay, Nemitz sued, and eventually the case was dismissed by a federal court. But as I mentioned earlier, private use and ownership of space bodies are still legally murky issues. So who knows? Maybe someday NASA will have to pay up.

There were other unmanned asteroid probes lofted into space as well—and not all were sent by NASA. For example, in 2004 the European Space Agency launched the *Rosetta* space probe to fly by main belt asteroid 21 Lutetia—an 80-mile-wide whopper whose surface is scarred with huge impact craters, some up to 30 miles wide—and asteroid 2867 Šteins, a much smaller diamond-shaped rock. (After scoping out the asteroids, the *Rosetta* spacecraft in 2014 dropped a lander module onto Comet 67P/Churyumov-

Gerasimenko, a raggedy, dumbbell-shaped comet that zips around the Sun every six years or so. Although there had been previous flybys of comets, this was the first time a man-made device had ever actually landed on one.)

Japan was in on the asteroid exploration game as well. In 2005 the Japan Aerospace Exploration Agency sent the *Hayabusa* spacecraft—the name means "Peregrine Falcon"—to Near-Earth asteroid 25143 Itokawa to collect samples and return them to Earth. Although the mission was plagued with technical problems, *Hayabusa* managed to land on the 200-yard-wide asteroid, pick up some tiny grains of surface material and fly back to Earth, where it parachuted a canister of the asteroid dust into the Australian Outback before the spacecraft burned up in the atmosphere. Undeterred, the Japanese space agency in 2014 sent the *Hayabusa2* spacecraft to Near-Earth asteroid 162173 Ryugu, a half-mile-wide, sugar-cube-shaped object rich in nickel-iron and other metals— about $85 billion worth, according to some estimates, which could make it a target for future asteroid-mining ventures. As I write this, in 2018, *Hayabusa2* is orbiting the asteroid, with plans to gather samples and return them to Earth in 2020.

The 1990s and early 2000s also saw a boom in terrestrial Near-Earth Object search programs. Tom Gehrels and Bob McMillan's ongoing Spacewatch survey on Kitt Peak in Arizona continued in operation, primarily providing follow-up NEO observations that are critical in determining an asteroid's orbit—and its chances of striking Earth. (Gehrels, a legendary figure in the planetary science world, died in 2011 at age eighty-six.) Gene Shoemaker's former associate, the redoubtable Eleanor "Glo" Helin, also remained in the NEO search business. In 1995 Helin and NASA's Jet Propulsion Laboratory launched the Near-Earth Asteroid Tracking program (NEAT) at Palomar Observatory in California and on a mountaintop in Hawaii, a program that discovered tens of thousands of asteroids and other space objects before it was discontinued in 2007. Helin, feisty as ever—her own husband described her as passionate about her work but "not an easy person"—died in 2009.

Some old-timers in the world of stargazing also got into the new NEO search game. For example, the venerable Lowell Observatory in Flagstaff, founded by wealthy Bostonian Percival Lowell back in 1894—Lowell was famous, or infamous, for claiming that he'd spotted artificial water-transporting "canals" on Mars—set up the NASA-funded Lowell Observatory Near-Earth Object Search (LONEOS) in 1993, and over the next fifteen years it racked up hundreds of Near-Earth Object discoveries. (Despite its contributions to NEO research, the Lowell Observatory is probably best known as the site from which self-taught astronomer Clyde Tombaugh discovered Pluto in 1930. Pluto may not be an official planet anymore—at least not according to the International Astronomical Union, which downgraded it to a "dwarf planet" in 2006—but the Lowell Observatory still proudly bills itself as "The Home of Pluto.")

There were new kids on the NEO search block as well. In 1998 Steve Larson and other astronomers at the University of Arizona Lunar and Planetary Laboratory founded the Catalina Sky Survey, which eventually became the most prolific NEO search program ever. (More on the Catalina Sky Survey in the next chapter.) Also in 1998, NASA, the U.S. Air Force and the Massachusetts Institute of Technology set up the Lincoln Near-Earth Asteroid Research (LINEAR) project at the vast White Sands Missile Range in New Mexico, not far from the Trinity atomic bomb test site. LINEAR took two Air Force telescopes originally designed to track man-made space objects such as satellites and turned them into NEO detectors—another indication that the U.S. government was finally taking the NEO threat to heart.

Interplanetary radar systems also got into the asteroid act, gathering data on hundreds of Near-Earth asteroids. For example, the giant radio telescope at Arecibo Observatory in Puerto Rico—its receiving dish is a thousand feet wide, the length of three football fields—can bounce signals off of asteroids millions of miles away and provide exact details on their size, shape, and surface features, something that's hard to do with ordinary optical telescopes. In 2017 Arecibo picked up detailed images of Near-Earth asteroid 3200 Phaethon, a three-mile-wide, egg-shaped rock that came within six million miles of our planet—pretty close by astronomical standards. Because of its size Phaethon is classified as a "potentially hazardous asteroid," and in 2093 it's scheduled to come even closer to Earth, within about two million miles. It will almost certainly miss us, but still, the more we know about asteroids like Phaethon the better—and radio telescope imagery plays a big role in that.

(The Arecibo facility is so large and visually striking that it's been used as a backdrop in a number of movies, including the 1995 James Bond movie *GoldenEye* and the 1997 film *Contact*, about the discovery of extraterrestrial beings. In fact, Arecibo famously played a real-life role in the search for extraterrestrial intelligence. In 1974 Arecibo beamed out a three-minute message containing a numerically coded graphic of the Solar System, a stick-figure depiction of a human, the atomic numbers for various elements and so on. The message was aimed at globular star cluster M13, a collection of hundreds of thousands of stars, one of which just might have a planet that supports intelligent life—ideally intelligent enough to have the wherewithal to figure out the message. Unfortunately, M13 is 22,000 light-years away—that's the distance that light travels in twenty-five millennia—so it will be a while before we hear back.)

Today there are other, more recent additions to the asteroid hunting effort as well. The University of Hawaii's NASA-funded

Panoramic Survey Telescope and Rapid Response System (Pan-STARRS) operates two NEO-detection telescopes atop the barren, 10,000-foot peak of Haleakala on Maui; since 2010 it has discovered thousands of Near-Earth Objects. In 2017 the Asteroid Terrestrial-impact Last Alert System (ATLAS)—also a NASA-funded University of Hawaii effort—became fully operational; it's designed to spot really small NEOs a few days or weeks before they impact. And there's another, even more powerful Near-Earth Object search system on the way in the next few years. It's called the Large Synoptic Survey Telescope (LSST), a $600 million facility currently under construction on a 9,000-foot peak in Chile. Actually the "large" in the name doesn't quite cover it; this telescope's reflecting mirror is *twenty-seven feet* in diameter, which means the telescope can cover vast swaths of the sky in a short amount of time; compared with small telescopes, it's like the difference between looking at the starry cosmos through a straw and looking at it through a picture window. In fact, the LSST will be able to search the entire visible sky every few nights, dramatically enhancing the rate of NEO discoveries.

Of course, Earth-bound asteroid searches have their limitations. With optical telescopes you can only search for asteroids at night—their reflected light from the Sun is too dim to spot during the daytime—and when the Moon is full and bright it has the same obscuring effect. Also, only a portion of the night sky is visible from any given geographic location at any given time, and sometimes clouds and weather can block the view of the sky entirely. All of those factors leave gaps in the sky coverage, gaps through which potentially hazardous asteroids can slip by undetected. But you can get around those obstacles by searching for Near-Earth Objects from space.

We're doing just that with a NASA-Caltech program called NEOWISE. The "WISE" part of the name stands for "Wide-field

Infrared Survey Explorer," which is a satellite-mounted telescope parked in Earth orbit. Originally, the 2009 WISE mission was designed to gather data on stars and distant galaxies, but in 2011 it was re-tasked to search for Near-Earth Objects—hence the "NEO" in the name. NEOWISE uses infrared thermal imagery, meaning it measures the heat glow of an asteroid, not its reflected light, which allows it to more accurately determine an asteroid's size. It's configured so that it always has its back to the Sun, which eliminates the daylight problem, and obviously there's no weather in space to mess up the view. Over the past seven years NEOWISE has examined tens of thousands of known asteroids and discovered hundreds of new NEOs. Unfortunately, poor old NEOWISE is already past its expected mission life, so NASA would like to replace it with a new infrared space telescope system called NEOCam—"Near-Earth Object Camera." But with a $500 million price tag it's not at all certain that NEOCam will get the necessary funding.

These are just a few of the ongoing efforts to find and track potentially hazardous asteroids. Although the U.S.-based sky surveys have discovered about 95 percent of the currently known NEOs, there are also NEO search programs in Canada, Japan, the European Union, the United Kingdom, and elsewhere.

The point is that we have vastly more eyes on the sky than we did twenty or thirty years ago. While the rest of us are sleeping, there's a new generation of asteroid seekers out there searching for Near-Earth Objects that could pose a danger to our planet.

And exactly how do they do it? How in the vastness of space can they detect such relatively tiny and obscure objects?

We can find the answer to that question on a mountaintop in Arizona.

CHAPTER 6

ASTEROID HUNTERS

It's a bitterly cold and satisfyingly clear March night at an old Air Force radar station perched atop the 9,000-foot peak of Mount Lemmon, in the Catalina Mountains northeast of Tucson. I'm sitting in a modest enclosure, not much more than a shed really, watching as a gray-haired, gray-bearded, 54-year-old astronomer named Richard Kowalski stares at a bank of flat-screen monitors connected to a giant telescope that's aimed at the starry sky.

Kowalski is a professional asteroid hunter. And on this night, as every night, asteroids are popping up everywhere.

"There's one," Kowalski says, mostly to himself, as he looks at computer images that show a small, blurry gray blot that seems to jump across the background of stars. The blot is an asteroid, most likely a previously unknown Near-Earth Object, so Kowalski punches in some commands on his keyboard to begin the process of identifying and tracking it. A few minutes later he spots another jumping blob. "Yep, that's a rock," he says, "probably a main belt object." Within seconds the computer tells him that the "rock" is indeed a known main belt asteroid, safely tucked away far from Earth and not a threat to our planet. Kowalski makes a note of it and then it's back to the images on the monitor again.

Over the course of this night Kowalski will look at thousands of computer images, searching for space rocks that one day could crash into Earth.

For Kowalski this is all routine, but for me it's pretty exciting—although I'm mindful that my presence here as a journalist is predicated on three requirements: First, I need to stay out of the way; second, I shouldn't ask too many dumb questions; and third, I must never, ever touch the red "hotline" phone! I can report that I more or less succeeded at requirement number one, and failed miserably at requirement number two—although Kowalski couldn't have been nicer or more patient about fielding my questions, dumb ones included.

As for requirement number three—not touching the red phone—well, that's just a bit of asteroid-hunter humor. There is a red push-button phone on Kowalski's desk, but it's just an ordinary landline, not a direct connection to the National Command Authority in case of an incoming Earth-buster asteroid. Kowalski explains that they used to have a beige phone, but "We switched to a red one so when people come up to visit we can tell them, 'Yeah, that's the hotline, direct to the president.'" You might not think that astronomers as a group have a great sense of humor, but in my experience most of them do. Kowalski, a bear of a man with big hands and a bigger laugh, certainly does.

Still, when it comes to his job Kowalski is deadly serious. Like others in the asteroid-hunting business, he knows that the chances of spotting an asteroid on a certain collision course with Earth are small. But the consequences of *not* detecting a dangerous asteroid, even a relatively small one, could be catastrophic.

"We have seven and a half billion people depending on us to do our jobs correctly," Kowalski says. "What if we missed something and then someday it comes down on a city? That would be hard. So we have to take it seriously."

Kowalski is part of the aforementioned Catalina Sky Survey, a NASA-funded Near-Earth Object detection program that grew out of the "Spaceguard Goal" of the late 1990s to find and track at least 90 percent of all large NEOs. Working under the auspices of the University of Arizona's Lunar and Planetary Laboratory, the survey operates two telescopes on Mount Lemmon—a 60-inch reflector for finding NEOs and a remotely operated 40-inch reflector for follow-up observations—as well as a 30-inch wide-field Schmidt telescope on nearby Mount Bigelow.

As a "senior research specialist," Kowalski is one of a half-dozen people in the Catalina survey—known as "observers"—who spend their nights searching for asteroids, backed up by another half-dozen engineers and technicians. The observers alternate manning the 60-inch telescope at the Mount Lemmon Observatory for about twenty-four nights each month, pausing only when the full or near-full Moon makes the sky too bright for what astronomers call "good seeing."

As noted above, the Mount Lemmon Observatory is situated on the site of a former Air Force radar-tracking installation, and although the Air Force personnel are long gone the place still has a distinct Cold War military feel. There are some old radar towers, some military barracks used by visiting astronomers and students, a recreation center with a pool table and basketball half-court, all scattered about on forty acres of mostly bare ground surrounded by chain-link fences. Just getting up there is an effort. As the crow flies the observatory is only a half-dozen miles away from the outskirts of Tucson, but to reach it requires a 26-mile climb up a two-lane paved scenic road with endless switchbacks and long ascending loops; getting back down, in the dark, and especially in sketchy winter weather, can be a white-knuckle experience for the uninitiated.

The 60-inch telescope at Mount Lemmon is housed in a 30-foot-diameter white metal dome with a section of retractable roof; the top of the dome can rotate around on rollers to face any portion of the visible sky. The telescope itself is huge, some twenty feet long, mounted on a yoke to give it vertical maneuverability. It's not the sort of long, thin tube we envision when we think of telescopes; instead it's a bulky device that's shaped sort of like a water heater.

Officially the telescope is described on the Catalina Sky Survey website as "an f/1.6 reflector equipped with a 111-megapixel (10,560 × 10,560 pixel) CCD detector mounted at prime focus. The field of view is 5.0 deg^2 with a pixel scale of 0.77"/pixel (unbinned)." What that means in English is that it's a massive light-gathering machine—Kowalski calls it a "light bucket"—that's capable of detecting the reflected light of space objects that are a billion times more faint than can be seen from Earth with the naked eye; it's sort of like standing in New York and being able to see a match lighted in Los Angeles. For comparison, when Tom Gehrels first started using CCD detectors for the Spacewatch sky survey in the early 1980s his CCD detector had a capacity of about 165,000 pixels; the Catalina 60-inch telescope CCD has a capacity of more than a *hundred million* pixels. It's the difference between one of those old 1980s cell phones that looked like a shoebox and a brand-new iPhone X.

If you're thinking of a working astronomical observatory like this one as an antiseptically clean place, manned by scientists in white lab coats looking through the eyepiece of a telescope, don't. Although the telescope itself gleams, the inside of the observatory dome is a little reminiscent of a machine shop. There's a toolbox and a workbench and wrenches hanging on the walls, old bolts and spare metal parts are piled up in cardboard boxes, and there's a scent of lubricating oil on mechanical gears. And professional astronomers don't actually plant their eyeballs on telescope eyepieces anymore, at least not at work; these days it's all done with cameras and computer images. As for Kowalski, no white lab coat for him. On this particular March night he's wearing hiking boots, a red down jacket, a gray sweatshirt and blue jeans; however, he is wearing white socks.

Once he has the retractable roof section open and the 60-inch ready to go—he calls the telescope "my partner in crime"—Kowalski repairs to the "control room" that's attached to the dome. Unlike the dome itself, the control room is heated, which is an important consideration at 9,000-plus-feet elevation, especially in the winter. It's a modest space, with government-issue beige linoleum floors and beige walls; the only attempts at adornment are a couple of posters portraying "The Threat and Lure of Near-Earth Objects" and the hazards to astronomy of urban light pollution. There's a bathroom and a small kitchen with an ancient refrigerator and stove where the observers can prepare their meals—Kowalski's menu tonight is split-pea-with-ham soup and a chicken chile verde burrito—and an even smaller room with a bare mattress where observers can rack out if they don't feel like driving back down the mountain in the morning.

Kowalski's action station is a rolling desk chair at a Formica table that holds a half-dozen flat-panel monitor screens. Wires and cables snake about, and there's the red push-button phone sitting there for landline calls that—it bears repeating—are not to POTUS. If anything really urgent comes up, NASA will be alerted and they'll take it from there. There's a coffee thermos to help him through the night, and when he's alone he'll sometimes listen to 1960s and '70s rock; Pink Floyd and a Long Island band called Zebra are particular favorites. Except for an occasional ten-minute break, Kowalski will spend all night in that chair, from sunset to sunrise, monitoring the images on the screens.

There are a lot of images to monitor. Every night the telescope and the CCD camera take four images each of some 250 relatively small patches of sky, called "fields," with the images taken some ten minutes apart. The computer will electronically assess those images to see if there's anything unusual there—say, a blob of light that seems to be moving in relation to the fixed-position blobs of light that are the stars. That could be an asteroid, and if it's a real fast mover it may be a Near-Earth Object.

But maybe not. The computer program has parameters to sort out unusual objects amid the cluttered "noise" of background stars. But what the computer lacks is *judgment*. It can't know for certain if a certain blob of light is a real space object or just some light anomaly, a visual trick of reflection or cosmic rays or an overheated pixel. So it flags the unusual objects and displays the images on the flat-screen monitor for what's known as "validation" by a human observer. It's the human observer who has to decide if it's a real object or just a false detection by the computer—and that's where Kowalski comes in.

"The software can't tell if it's real or not," Kowalski tells me. "But the human eye and brain are great at pattern recognition; we can detect a real object in the noise that a computer can't."

Mind you, what Kowalski sees on the screen aren't finely detailed images of potato-shaped asteroids zipping along like the ones in *Star Wars*. They're just small blobs of light. So deciding which ones are real asteroids and which are not is part science and part art.

For Kowalski it's a fast-moving game of "yeses" and "nos." Is the gray blob on the screen jumping around in different directions? Is it visible in one image but not in another image taken ten minutes later? Then it's not a real asteroid, in which case Kowalski moves on to the next set of images. The principle at work here is that it's better for the computer program to churn out a lot of false detections than to miss even one real Near-Earth Object. So on any given night Kowalski will examine eight to ten thousand potential asteroid detections coughed up by the computer, and in the vast majority of cases they turn out not to be asteroids.

But every so often Kowalski takes a look at an object that's been flagged by the computer as a potential asteroid and in the briefest glance he determines that, yes, it's an asteroid all right. At that point Kowalski checks to see if the object is a known asteroid, one that's already been discovered; if so, he'll take note of its position and then move on. If it's a previously *un*known object, a new discovery, Kowalski will decide if it's a main belt asteroid, one that's safely locked in orbit way out between Mars and Jupiter. If it's an undiscovered main belt-er, okay, that's kind of cool, but nothing to get excited about; it's not going to come anywhere near Earth. Kowalski will note its astrometric location and open a file on it.

But what Kowalski and the Catalina Sky Survey are really looking for are the undiscovered Near-Earth Objects, the ones that could someday pose a hazard to our planet. If Kowalski decides the blob on the flat screen is a new Near-Earth Object, it sets in motion an elaborate confirmation and tracking process. Kowalski notes the potential NEO's position and brightness and sends an email report to the International Astronomical Union's Minor Planet Center (MPC) at the Harvard-Smithsonian Center for Astrophysics in Cambridge, Massachusetts, the worldwide clearinghouse for data on asteroids and comets. The MPC computer will confirm that it's a new discovery and roughly calculate the NEO's orbit. Then it posts that information on its "NEO Confirmation Page," a publicly accessible Internet site, so that other astronomers around the world—many of them amateurs—can do follow-up observations over the next few days. (Amateurs play an important role in the process; more on that in a moment.) If it's a particularly interesting object, Kowalski will schedule an initial follow-up observation with the Catalina Sky Survey's own telescopes as well. The follow-ups are important, because the more observations you get, the easier it is to work out the asteroid's precise orbit—its identifying signature in space—and determine if the asteroid poses a future threat to Earth.

Meanwhile, the data on the newly discovered NEO are also monitored by NASA's Center for Near-Earth Object Studies at the Jet Propulsion Laboratory (JPL) in Pasadena, California, where computers calculate what the chances are that the asteroid will pass very close to or collide with Earth—not just today or tomorrow but even a century into the future. (A European service called NEODyS—Near Earth Objects Dynamic Site—provides the same function.) If a potentially hazardous asteroid has even a tiny chance of hitting Earth within the next century it's posted on JPL's publicly available "Sentry Risk Table." The asteroid also gets a "hazard rating" based on what's known as the Torino Scale.

The Torino Scale defines the impact hazard of a given asteroid on a scale of "0" to "10." A zero rating means there's virtually no chance that the asteroid will hit Earth within the next century. A "3" means there's a 1 percent or slightly better chance the asteroid will hit Earth sometime within the next century and cause "localized destruction"—as in possibly taking out a city. A "6" means "a serious but still uncertain threat of a global catastrophe"— astronomers will have to keep a close eye on the thing—while an "8" means a definite hit by an asteroid big enough to wipe out Kansas City. Both the Meteor Crater impact and the Tunguska Event would have rated "8" on the Torino Scale.

And a "10"? A "10" is a kiss-your-ass-goodbye moment. (Of course, that's my language, not the Torino Scale's.) Officially a "10" means there's a 100 percent chance the asteroid will hit Earth with enough force to cause a "global climatic catastrophe that may threaten the future of civilization as we know it." A "10" level impact is expected to occur no more often than once every hundred thousand years or longer, but again, that's a probability, not a prediction.

It's important to note that the highest Torino Scale hazard rating ever applied to an asteroid was a "4." That dubious honor was briefly awarded to the 400-yard-wide asteroid called Apophis, named after an ancient Egyptian god of darkness. After its discovery in 2004 Apophis was projected to make a close approach to Earth in 2029—on a Friday the 13th no less—and it was initially given a 3 percent chance of actually hitting Earth. If it did hit Earth it would pack the punch of 750 megatons of TNT, about fifteen times more powerful than the biggest nuclear bomb ever detonated, so even a 3 percent chance caused something of a stir. But Apophis's "4" rating was soon downgraded to "0" after more observations of the asteroid's orbit came in, although it's still near the head of the line in the perp walk of Earth-threatening asteroids. On that Friday the 13th in 2029 Apophis will pass within 19,000 miles of Earth, closer than geosynchronous communication satellites, but the guys at NASA are confident it will miss us.

The same process applies with other newly discovered NEOs. Occasionally one might rate a "1" or a "2" when it's first observed, but as more observations come in it's usually downgraded to a "0" or taken off the risk table altogether. As it now stands, the Center for Near-Earth Object Studies at JPL maintains a publicly available list of hundreds of close-approaching Near-Earth asteroids, none of which has a current Torino Scale rating higher than "0"—which means no significant known asteroid is likely to hit Earth in the next century. But remember, that only refers to the asteroids we already know about; as we'll see, there are still a lot of undiscovered NEOs out there. And there's always a chance—a small chance, but a chance nonetheless—that tomorrow an asteroid-on-asteroid collision out in the main belt could suddenly send our way a previously harmless asteroid. Space is full of surprises.

Anyway, that's what happens when Kowalski or another observer spots an NEO—and it happens pretty often. On this night Kowalski will spot two previously undiscovered Near-Earth Objects that are big enough—about 150 to 300 yards wide—and come close enough to Earth to qualify as potentially hazardous objects. Most nights Kowalski or other Catalina observers will spot about eight to ten NEOs; Kowalski's personal best is twenty-one NEOs detected in a single night of observing.

How many asteroids has Kowalski personally discovered in his entire career? Actually, he can't remember.

"Gosh, I have no idea," he tells me. "There've been many thousands of main belt asteroids, many hundreds of NEOs, so they all kind of run together. It's not that I'm jaded or anything. It's neat when you discover a main belt asteroid that no one on Earth has ever seen before. But remember, there are tens of millions of those pieces of rock out there."

Still, there have been some memorable discoveries—memorable not only to Kowalski but to the world of astronomy in general. In fact, although he modestly shrugs it off, Kowalski is something of a celebrity within the asteroid-hunting community. Much of that has to do with a small piece of space rock called 2008 TC_3.

On October 6, 2008, Kowalski was on duty with the 60-inch telescope on Mount Lemmon when he spotted what appeared to be a small undiscovered Near-Earth asteroid—it turned out to be about thirteen feet wide—which he duly reported to the Minor Planet Center as described above. There was nothing outwardly unusual about this particular NEO, one of several he had discovered that night; it was just another routine discovery in a routine night of hunting asteroids. At the end of his shift that morning Kowalski packed up and went to bed.

But while Kowalski was sleeping, the rest of the asteroid-hunting community was going crazy. When the then-director of the Minor

Planet Center, Tim Spahr, checked the overnight reports of asteroid observations on the MPC computer, the computer showed that Kowalski's asteroid, now designated 2008 TC$_3$, was going to hit Earth—and not in a year or a decade or a century. Asteroid 2008 TC$_3$ was going to hit the Earth that very night! Spahr announced that astonishing news on a series of publicly available "Minor Planet Electronic Circulars," and soon scores of professional and amateur astronomers around the world were frantically manning their telescopes and sending in reports about the asteroid's position. Hundreds of reports came in from Russia, Australia, the United Kingdom, the Czech Republic, even from an amateur observatory on the campus of New Milford High School in Connecticut. What made it so exciting was that this was the first time ever in the history of the world that an incoming asteroid had been spotted *before* it collided with Earth. It was an unprecedented opportunity for scientists to study an asteroid-Earth impact in real time.

At the same time, the guys at the Jet Propulsion Laboratory not only confirmed that the asteroid was indeed going to strike Earth, but they also figured out precisely where and when. Asteroid 2008 TC$_3$ was going to pass over Africa on a southwesterly course and enter the atmosphere over the Nubian Desert in Sudan at 5:46 a.m. local time.

"That was an interesting day," recalls Paul Chodas, manager of the Center for Near-Earth Objects Studies program at JPL. Working with JPL astrodynamicist Steve Chesley, Chodas calculated out the impact point within a kilometer, then checked out the location on Google Earth. The nearest inhabited place was miles away—population ten.

Because of the asteroid's small size nobody was really worried about any damage from the impact. The multiple observations had determined that the small rocky asteroid would almost certainly explode harmlessly high in the atmosphere. Still, the Minor Planet

Center alerted a senior NASA official, and just to be on the safe side, NASA sent out alerts to the White House, the Pentagon, and other national security agencies.

Meanwhile, Kowalski was still sleeping. When he woke up that afternoon some of his co-workers told him that the asteroid was going to hit the Earth in a few hours, but he thought they were pulling his leg. It was only when he checked his email and found hundreds of messages congratulating him on his discovery that he realized it wasn't a joke.

"I was, literally, the last guy in the [asteroid-observing] community to know about the impact," he says, amused at the thought.

2008 TC_3 performed precisely as predicted, blowing up high over the Nubian Desert in a brilliant fireball that was seen from hundreds of miles away; the explosion had the energy equivalent of about a kiloton. Later an American astronomer named Peter Jenniskens teamed up with University of Khartoum professor Muawia Shaddad to mount a search for pieces of the asteroid. A busload of Sudanese university students lined up side-by-side and started marching across the desert like police cadets searching for a body, and eventually they came up with hundreds of small pieces that together weighed about twenty-four pounds. Later Kowalski was given one of those small pieces in honor of having discovered the asteroid.

The discovery of 2008 TC_3 alone would have been enough to put Kowalski in the astronomy record books. But just after midnight on New Year's Day in 2014—asteroid hunters don't stop for holidays—Kowalski was up on Mount Lemmon when he spotted another incoming asteroid, designated 2014 AA. It was about the same size as the earlier 2008 TC_3, but this one exploded over the Atlantic Ocean about twenty hours after he discovered it. And then—this is hard to believe—on the night of June 2, 2018, Kowalski spotted *another* incoming asteroid,

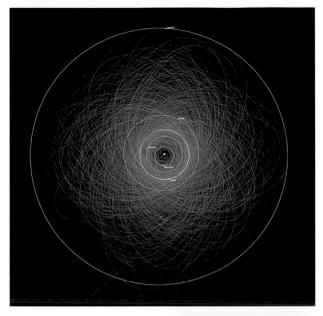

The Solar System is a busy place. This graphic depicts the known orbits of 1,400 asteroids that could someday pose a threat to Earth. Many more have yet to be discovered. (NASA/JPL-Caltech)

An artist's conception of an asteroid hurtling toward Earth. Small asteroids usually explode or burn up when they hit the atmosphere. Larger ones can strike Earth's surface with the force of multiple nuclear bombs. (Oliver Denker/Shutterstock)

Asteroids come in an endless variety of shapes. Near-Earth asteroid 433 Eros is ten miles wide and shaped like a shoe. (NASA/JPL/Johns Hopkins University Applied Physics Laboratory)

An artist's conception of the so-called Halloween asteroid, based on radar imagery. From some angles the 2,000-foot-wide asteroid looks chillingly like a human skull.
(J. A. Peñas/SINC [Servicio de Información y Noticias Científicas])

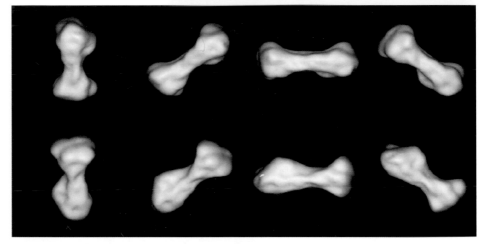

Computer generated images of Main Belt asteroid 216 Kleopatra, based on radar observations. Kleopatra, which resembles a very large dog bone, is roughly the size of New Jersey.
(Stephen Ostro, JPL/NASA)

An artist's conception of asteroid 16 Psyche, a giant 120-mile-wide piece of space metal. It's believed to be the iron core of a larger body that was broken apart.
(SSL/ASU/P. Rubin/NASA/JPL/Caltech)

Italian astronomer Giuseppe Piazzi discovered the first known asteroid, Ceres, in 1801 while searching for a missing planet between Mars and Jupiter. (Painting by Costanzo Angelini [1760–1853]/Wikimedia)

Meteor Crater, Arizona. Almost a mile wide and hundreds of feet deep, this crater was formed fifty thousand years ago when an asteroid struck the Earth with the explosive energy of a thousand Hiroshima bombs. (Copyright Greg McKelvey, www.gempressphotos.com, used with permission.)

Most of Earth's impact craters have eroded away. Because of its young age and desert location, Meteor Crater is the best preserved. (Copyright Greg McKelvey, www.gempressphotos.com, used with permission.)

In 1902, wealthy Philadelphia mining engineer Daniel Moreau Barringer hatched a scheme to dig up the asteroid that formed Meteor Crater and harvest its valuable minerals. Barringer spent most of his fortune on the scheme, only to learn that the asteroid no longer existed. (The Barringer Crater Company)

Private companies plan to mine asteroids in space for fuel-producing water and valuable minerals. This is an artist's conception of how a solar-powered robotic vehicle would lock onto a small asteroid and extract the ore. (Bryan Versteeg/Deep Space Industries)

The atmospheric explosion of an asteroid or comet in the remote Tunguska region of Siberia in 1908 was the biggest known cosmic impact in modern times, devastating a 100-square-mile area and knocking down millions of trees. Russian scientist Leonid Kulik mounted several arduous expeditions to the scene. Kulik and the Tunguska Event were memorialized in this 1958 Soviet postage stamp. (Wikimedia Commons)

Astrogeologist Gene Shoemaker was among the first to warn that devastating asteroid impacts on Earth aren't just possible—they're inevitable. (U.S. Geological Survey)

Gene Shoemaker, his wife Carolyn, and astronomer David H. Levy discovered Comet Shoemaker-Levy 9. This composite of images shows fragments of the comet before they slammed into Jupiter in 1994. (The round black spot on Jupiter's surface is the shadow of Jupiter's inner moon Io.) (NASA, ESA, H. Weaver and E. Smith (STScI) and J. Trauger and R. Evans [NASA/JPL])

Geologist Walter Alvarez (right) and his father, Nobel Prize–winning physicist Luis Alvarez, rocked the scientific establishment with their theory that the impact of a six-mile-wide asteroid wiped out the dinosaurs. Walter Alvarez has his right hand on a thin layer of iridium-rich clay in an Italian gorge that helped prove the theory. (Lawrence Berkeley National Laboratory)

Death comes to the Cretaceous: an artist's conception of the dinosaurs' worst day. (Esteban De Armas/Shutterstock)

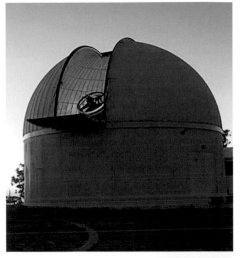

The Catalina Sky Survey, funded by NASA and operated by the University of Arizona Lunar and Planetary Laboratory, has discovered more than nine thousand Near-Earth asteroids—more than any other NEO search program. CSS facilities include a 60-inch telescope at the observatory atop the 9,200-foot peak of Mount Lemmon outside of Tucson. (Catalina Sky Survey, used with permission.)

Richard Kowalski of the Catalina Sky Survey at the observatory on Mount Lemmon (see above). To date only three small asteroids have been discovered before they hit Earth—and Kowalski discovered all of them. (Gordon L. Dillow)

This is what a small incoming asteroid looks like in a telescope image. Richard Kowalski discovered asteroid 2014 AA (in pink circle) twenty hours before it exploded over the Atlantic Ocean. The surrounding bright lights are distant stars. (NASA/JPL-Caltech/Catalina Sky Survey-University of Arizona)

On February 15, 2013, a 20-yard-wide asteroid blazed through the atmosphere and exploded in the sky near the Russian city of Chelyabinsk, sending a powerful shock wave to the ground. More than 1,500 people were injured, mostly by flying glass. (Aleksandr Ivanov/Wikimedia Commons)

Lindley Johnson, America's Planetary Defense Officer, at a 2016 exercise conducted by NASA and FEMA. Johnson's Planetary Defense Coordination Office at NASA develops plans to deal with incoming asteroids. (The Aerospace Corporation)

NASA's unmanned OSIRIS-REx mission will collect surface samples from potentially hazardous asteroid 101955 Bennu and return them to Earth in 2023. (NASA)

designated 2018 LA, that was 240,000 miles away from Earth and coming in fast at 38,000 miles per hour. The tiny, six-foot-wide asteroid burned up in the atmosphere over the southern African nation of Botswana just a few hours later. Bear in mind, in all of human history only three asteroids have ever been spotted *before* they hit Earth—and Kowalski discovered all three of them. It almost makes you wonder if the guy is possessed of some asteroid-detecting superpower.

Not at all, Kowalski says. Just simple luck of the draw.

"I was just lucky enough to be sitting in the [observer's] chair at the right times," Kowalski tells me, "and I just happened to point the telescope at the right patches of sky. Any of our observers could have done the same thing."

There are other discoveries on Kowalski's record. In 2011 he discovered a tiny, one-yard-wide asteroid called 2011 CQ_1 that passed within 3,400 miles of Earth—the closest-ever approach of an asteroid that didn't hit the Earth. He has also discovered more than a dozen comets that bear his name, including the re-discovery of Comet Pigott-LINEAR-Kowalski, which was first observed way back in 1783 and had been lost ever since.

(Other observers with the Catalina Sky Survey have spotted some unusual NEOs as well. For example, in 2006 Catalina observers picked up a tiny, six-feet-wide asteroid called 2006 RH_{120} that had been temporarily "captured" into Earth orbit, circling the Earth four times at long range before zipping back into solar orbit; it's due back in 2028. In 2015 Catalina observer Rose Matheny detected WT1190F, an NEO that was about to hit Earth—but this NEO wasn't an asteroid. It was a piece of space junk, probably a portion of some long-ago rocket booster; it later spectacularly burned up in the atmosphere over the ocean near Sri Lanka. And remember that cherry-red Tesla Roadster with a spaceman mannequin inside it that Elon Musk and the private space exploration firm SpaceX launched into solar orbit in early 2018? Catalina observers picked that up, too—although again, it showed up on the telescopes as only a tiny blob of light. By the way, calculations show that the Tesla has a 6 percent chance of T-boning the Earth in the next three million years.)

Given all his scientific achievements, it may surprise you to learn that while Kowalski is a professional astronomer, he doesn't have a university astronomy diploma—or any college diploma—hanging on his wall. Born and raised on Long Island, Kowalski was one of those guys who wanted to do things, not sit in a classroom and learn about them. His interest in astronomy started early, when his father put up a giant map of the Moon so they could track the Apollo lunar landings. That led to a small two-inch telescope in the backyard, then a six-inch one on top of the family garage. Kowalski was hooked.

After high school Kowalski got a job as an aircraft loader, eventually winding up with US Airways in Tampa; he also got a commercial pilot's license and worked as a flight instructor. When the aviation business went south after 9/11, he took a night job as a big-rig truck driver hauling Portland cement around Tampa. That was his work, but his life was always about the astronomy.

While in Florida, Kowalski built his own observatory in his backyard—he called it the "Quail Hollow Observatory" after the neighborhood he lived in—that consisted of a garden shed with a retractable roof and a seven-inch computer-driven CCD telescope. Originally he concentrated on the usual astronomy stuff—planets, stars, galaxies—but one day he read an article in a popular science magazine titled "Discover Asteroids from Your Backyard," and that was that. Finding asteroids became his passion. He discovered his first in 1998, a previously unknown four-mile-wide main belt asteroid, and while today he may not remember the thousands of other asteroids he's discovered, he remembers that one.

"I thought, Wow, I can actually discover asteroids!" Kowalski says. "That was really a thrill."

(Kowalski named that first asteroid 14627 Emilkowalski after his dad, Emil. Later he named another main belt asteroid 149951 Hildakowalski after his mom, Hilda. Kowalski also has an asteroid named after him—7392 Kowalski. It was discovered by astronomer Ted Bowell at the Lowell Observatory in Flagstaff, who named it after Kowalski in recognition of his contributions as an amateur astronomer.)

Technically Kowalski was an amateur, but although his telescope was less powerful than those used by professional sky surveys, his skill set matched those of any pro, and he was becoming increasingly well known in the professional asteroid community. In 1998 he founded the "Minor Planet Mailing List," an Internet information exchange site for both amateur and professional asteroid hunters that now has some two thousand members, and he became a leading advocate for the role of amateurs in Near-Earth Object detection.

As I've mentioned, while professional asteroid-detection programs discover most Near-Earth Objects, follow-up observations by amateurs can help to determine an NEO's precise orbit and approximate size—which are important things to know if an asteroid is zipping around close to Earth. Without the amateurs, the pros would have to spend precious big-telescope time doing their own follow-ups—which means they'd have less time to look for new potentially threatening asteroids.

"Our time is a lot cheaper than theirs is," says amateur astronomer Gary Hug, an ebullient 67-year-old retired machinist who uses a 22-inch telescope mounted in a retractable-roof shed in his backyard in Kansas; he calls it the "Sandlot Observatory." Over the years Hug has made close to twenty thousand follow-up observations of Near-Earth asteroids—and he has also discovered some new ones. He named his first asteroid after his wife—asteroid 15992 Cynthia—which Hug figured was only prudent, given the fact that he spent $10,000 on his self-constructed telescope and devotes most of his nights to it.

"That's the sort of thing you have to do when you're getting into something like this," Hug says, chuckling. Hug has received some small private grants to update his equipment from the Planetary Society's "Shoemaker NEO Awards" program. But he's never been paid a dime for the thousands of hours he's spent doing follow-ups on NEOs. So what's the motivation?

"It's fun," says Hug. "And the incredible thing is that you can contribute something that could possibly affect the entire Earth."

Kowalski made the switch from amateur to pro in 2005, when the Catalina Sky Survey was installing an improved telescope system and needed a skilled observer. Kowalski had missed out on other professional astronomy jobs because he lacked the required academic credentials, but Steve Larson, the survey's founder, put more stock in an observer's proven abilities than in a piece of paper in a frame. Kowalski got the job, and he's been there ever since.

For Kowalski and the other Catalina observers it's not always an easy gig. When he's observing, Kowalski generally works three nights on, three nights off. His working "day" starts just before sunset and ends just after sunrise. He eats lunch around midnight, supper at 8 a.m., and has his morning coffee in the early afternoon; sleeping is a constant battle with his circadian rhythms. Because the 28-day cycle of the Moon sets the observing schedule, he'll work Sunday-Monday-Tuesday nights one week, Friday-Saturday-Sunday nights another, which makes it tough for him and his wife, a small business owner, to plan a social life.

Don't get the wrong idea. Kowalski's not complaining. It's important and satisfying work, he doesn't really mind the hours—

"I've always been a night owl"—and it's peaceful up here on the mountain. It doesn't bother him at all that only a small percentage of people have any idea that he and other asteroid hunters are out there protecting them from potential cosmic impacts.

"Most people aren't aware of what we do," he says. "They don't know there are people searching for Near-Earth Objects. It's solitary work, but that's okay. It comes with the job."

Well, Kowalski may be a night owl, but I am not. Along about midnight I climb into my truck and with a final word of caution from Kowalski—"Don't forget to use your lower gears; it'll save your brakes"—I head downhill. A white-knuckle hour later I'm back in the city, where everybody is sleeping. Meanwhile, perhaps accompanied by the ethereal strains of Pink Floyd's "The Great Gig in the Sky," Kowalski is still up there on the mountaintop.

Hunting.

○

Again, the Catalina Sky Survey isn't the only Near-Earth Object search and track program out there. As I mentioned earlier, there's also Pan-STARRS, ATLAS, Spacewatch, NEOWISE, and others that are on the way.

Having a lot of NEO surveys operating at once isn't an unnecessary redundancy. Some of the surveys are better equipped to find certain kinds of asteroids than others, some concentrate on follow-up observations, and each survey is limited by its location to what part of the sky is visible on any given night. All of the active NEO surveys cooperate and share information, so it's a group effort—although they all keep a close eye on their individual stats. Currently the Catalina Sky Survey is the leader, having discovered almost half of all known NEOs over the past two decades. Although the newer Pan-STARRS in Hawaii briefly took the lead in yearly NEO discoveries, Catalina was back on top in 2017 with nearly a thousand new NEOs discovered.

"It's not really a competition," says Catalina Sky Survey director Eric Christensen, a tall, soft-spoken 40-year-old who's been in the asteroid hunting business for much of the past two decades. But after noting the disbelieving look on my face he smiles and cheerfully admits that, okay, yeah, it is a competition—"a friendly competition." It's like being in a softball league: You can like and respect the people on the other team, but you still want to score more runs than they do.

Collectively the various NEO sky surveys have racked up some impressive discovery numbers over the past two decades. In 1998 only 450 Near-Earth Objects had been discovered; as of October 2018, the number of known NEOs was at *19,000* and counting, with about 1,900 of them designated as potentially hazardous objects, meaning they're big enough and come close enough to Earth to pose a future threat. As for the non-Earth-threatening main belt asteroids, in 1998 fewer than 10,000 had been discovered; today about 750,000 such asteroids have been identified.

That's a lot of asteroids. But compared with the total number of asteroids out there—millions and millions of them, large and small—we've barely scratched the surface.

The asteroid hunters have made excellent progress in spotting the larger Near-Earth asteroids. You'll recall that in 1998 Congress directed NASA to identify at least 90 percent of all NEOs bigger than one kilometer (a little over half a mile) across, and to do it within a decade. The thinking was that those large asteroids (or comets) are capable of destroying civilization, if not human life itself, and thus pose the greatest danger.

The asteroid hunters have exceeded that goal. Today about 900 NEOs bigger than a half-mile across have been identified, representing about 95 percent of the estimated large NEO population.

And it turns out that none of those large, potentially Earth-busting asteroids are in orbits that will put them on a collision course with Earth, at least not within the next century and probably even longer. Of course, that still leaves 5 percent of the suspected large NEO population lurking out there, undiscovered, but it has dramatically reduced the known risk of a large asteroid impact anytime soon. As for the millions of small Near-Earth asteroids— from the size of a beach ball to the size of a small car—those don't pose a significant danger either. If they do ever hit us they'll burn up in the atmosphere.

So we're good, right? No worries? We've got this asteroid problem licked?

Well, not exactly. As Christensen explains it, "It's not the biggest ones we have to worry about; the chances of one of them hitting Earth are very small. And it's not the smallest ones we have to worry about; small asteroids hit us all the time. It's the ones in between we have to be concerned with."

And how many "in-between" Near-Earth asteroids are out there? Oh, maybe a million of them, ranging in size from about thirty yards wide up to half a mile wide. True, most of those are on the lower end in size, and the smaller the asteroid, the less danger it poses to Earth. But remember, the asteroid that blasted out the mile-wide Meteor Crater fifty thousand years ago was only about fifty or sixty yards wide, and yet it wiped out everything for several miles around. And the asteroid that leveled a hundred square miles of forest at Tunguska in 1908 was about the same size—about sixty yards wide. A small asteroid wouldn't wipe out human civilization, but it could do significant damage.

The problem is that spotting all of those small asteroids is a huge task. After directing NASA to search for half-mile-wide and larger Near-Earth asteroids in 1998, in 2005 Congress expanded the mission to identifying at least 90 percent of NEOs bigger than

150 yards wide by 2020. There are an estimated 25,000 NEOs of that size out there, but after two decades of asteroid surveys only about a third of those have been identified. And that still leaves hundreds of thousands of other potentially dangerous asteroids between 30 and 150 yards wide that we don't know about. We have no idea where they are, or where they're headed.

The point is that the asteroid hunters still have plenty of work to do. The question is whether we'll continue to give them the resources they need to do it.

In 2017 NASA spent about $60 million on Near-Earth Object tracking and planning programs. That was way up from the $20 million allocated in 2012, but it was only a tiny portion of NASA's $18 billion budget that year. And compared to, say, the Department of Defense's $700 billion budget, $60 million is chump change, a rounding error.

True, some NEO-tracking hardware can be expensive. For example, the aforementioned NEOCam space vehicle would cost about $500 million to design and launch, but once it's up there the annual operating costs would be minimal. NASA desperately needs that space-based NEO detection capability if it's ever going to finish finding and tracking all of the potentially hazardous NEOs, but whether the U.S. Congress and American taxpayers will be willing to pay the freight is an open question.

In fact, some taxpayers apparently think that even the small amount of money we currently spend on NEO programs is too much. In 2017, for example, I was at a university public forum on asteroids, led by a panel of experts who explained asteroid fundamentals and current asteroid detection programs to a largely non-scientist audience. When it comes time for Q&A, this young guy in the audience stands up and asks the same question that's asked of almost everybody who's involved in space exploration and study: "With all the problems we have here on Earth, why spend money on something like this? Don't we have more important things to worry about, like climate change?"

And suddenly the panelists start throwing sideways looks at one another, like, Okay, who's gonna have to field it this time? They've all heard the argument before, countless times. So finally one of the panelists gets up and explains that, yes, things like climate change are important, and yes, the chances of a major asteroid strike are low, but the amount of money spent on NEO detection and study programs is actually pretty small, and it could really help protect us in the future . . . and so on. The panelist was articulate, and knowledgeable, but he actually seemed a little defensive. With the zeal of the newly converted, I wanted him to shout out that we should be spending more on asteroid detection—and space programs in general—not less!

Catalina survey director Christensen understands the problem. He's low-key about the asteroid threat, and tries to keep it in perspective. The way he sees it, tracking potentially hazardous NEOs is like an insurance policy for future generations, assembling the knowledge they'll need if a truly civilization-ending impact is ever coming Earth's way. But like any insurance policy for a low-probability event, it can be a tough sell.

"Communicating the risk of asteroid impacts is difficult," he tells me. "The chance of a large asteroid hitting Earth and causing widespread destruction is low in any given human lifetime. But at the same time, there are a lot of smaller Near-Earth Objects out there that could potentially pose a threat."

Christensen smiles and adds, "It's the Chicken Little story, right? You don't want to oversell it—but then, someday the sky *is* going to fall."

Of course, funding for space projects is always a precarious proposition; just ask the astronauts who were slated to fly Apollo Moon missions 19 and 20, which were canceled due to waning

public interest. The good news on that score is that according to a
2018 Pew Research poll of 2,500 Americans, a solid majority—62
percent—said it was important for NASA to continue its search
for potentially dangerous asteroids, while only 9 percent said it
was a waste of time and money. By comparison, in 1993 only
25 percent of the public had even heard of the impact hazard.
(Although there was strong support for NEO detection programs,
the same Pew poll found that only small minorities—less than
20 percent—said that sending humans back to the Moon and to
Mars should be a top priority.)

So despite the occasional with-all-the-problems-we-have-on-
Earth grumbling from some members of the public and in Congress, it looks as if funding for NASA's NEO detection program
will continue, at least for the immediate future. That's partly
because of the sheer numbers of potentially hazardous NEOs
that Catalina and the other sky surveys have detected. Unlike in
the 1980s and early '90s, when most people didn't know we even
had an NEO problem, we now know the threat is there, even if
we don't yet know its full extent. And it's partly because, as I said,
space is full of surprises that tend to keep our interest up.

One of those surprises came in late 2017, when the guys at
Pan-STARRS spotted an asteroid that was truly out of this world.
Named 'Oumuamua (which roughly translates from Hawaiian as
"Scout"), the asteroid was 500 yards long and in an artist's ren-
dition appeared skinny and sharp-edged—it looked sort of like a
prison shiv—and it was hurtling past the Sun at a sizzling 196,000
miles an hour. What made it particularly unusual was that unlike
every other known asteroid, 'Oumuamua wasn't from our Solar
System. Instead, it had been tossed out of another star system
light-years away in space and by chance had happened to wander
through our Solar System—making it the first interstellar asteroid
ever discovered. Shortly thereafter, researchers revealed what may

be the second interstellar asteroid ever found—emphasis on the maybe; its origin hasn't been proven yet. Unlike the planets and most other bodies in the Solar System, this asteroid, designated 2015 BZ$_{509}$, revolves around the Sun *backwards*—that is, it moves clockwise, not counter-clockwise in its orbit, like a race car going the wrong way at the Indianapolis 500.

Those are just a couple of the recent surprises that the search for asteroids has turned up, and undoubtedly will continue to turn up in the future. There are a lot of strange things out there waiting to be discovered.

But perhaps the most dramatic and shocking asteroid surprise in recent years came along in 2013. And as shocking surprises tend to do, it came when the entire world was looking the other way.

CHAPTER 7

PLANETARY DEFENSE

In February of 2013 astronomers and space buffs around the world were eagerly awaiting the arrival of an asteroid dubbed 2012 DA_{14}.

Asteroid 2012 DA_{14} had been discovered a year earlier by a Spanish dental surgeon and amateur astronomer who was monitoring images taken from an observatory in Spain. In most respects it was a run-of-the-mill asteroid, a piece of rock about 150 feet wide, one of a million small NEOs flying around out there. But what was interesting about asteroid 2012 DA_{14} was that on February 15, 2013, it was going to pass within 17,000 miles of Earth—closer than many geosynchronous communications satellites.

There was no chance that 2012 DA_{14}—later named 367943 Duende after a mythological goblin-like creature—was going to strike Earth. It had been observed numerous times and its trajectory thoroughly plotted. And this wasn't the closest known approach ever by an asteroid. Two years earlier a little three-foot-wide asteroid named 2011 CQ_1—which had been discovered by Richard Kowalski—had zipped past Earth at a record-breaking distance of just 3,400 miles.

Still, 2012 DA_{14}'s flyby was the closest known approach by an asteroid of its size. And unlike little 2011 CQ_1, which had been spotted just hours before its close pass by Earth, astronomers had known for a year that 2012 DA_{14} was on its way, which gave them plenty of time to prepare. 2012 DA_{14} wasn't big enough to be seen from Earth with the naked eye, but in some parts of the world a backyard telescope or even a really good set of binoculars would be able to pick it up. Everybody in the asteroid community, professionals and amateurs alike, was excited.

The public was excited, too. The approach of 2012 DA_{14} prompted a flurry of newspaper headlines—"Asteroid to Buzz Earth," "Earth to Narrowly Escape Collision with Asteroid," and so on. NBCNews.com ran a story titled "Talk About Close! Asteroid to Give Earth Record-Setting Shave." Meanwhile, NASA was calling 2012 DA_{14}'s arrival "an Earth flyby reality check" in its press releases, a reminder that asteroids bear watching, and it had a panel of asteroid experts standing by to answer reporters' questions. And on the very day the asteroid was scheduled to pass by, a group of international astronomers was gathered at a United Nations–sponsored NEO conference in Vienna, using 2012 DA_{14}'s close pass to highlight the need for improved NEO tracking programs around the world.

So asteroid 2012 DA_{14} was a pretty big deal. But then, just sixteen hours before it was to make its closest approach, the word started to spread: Something strange had happened. In Russia. At this place called Chelyabinsk.

And suddenly, no one much cared about asteroid 2012 DA_{14} anymore.

Chelyabinsk is a sprawling, smoggy industrial city of some one million souls, situated about a thousand miles due east of Moscow. It does not have a particularly pretty history. During the Cold War the Chelyabinsk region was a center of Soviet military weapons production, including the Mayak plutonium-producing complex fifty miles northwest of the city, largely built by gulag slave labor. Nuclear waste from the facility was routinely dumped into the nearby Techa River, sickening hundreds of villagers, and in 1957 an explosion at Mayak released almost as much radiation as the Chernobyl nuclear power plant disaster in Ukraine in 1986. Although you won't see it in any Chamber of Commerce brochures, Chelyabinsk and its environs have been described as "the most radioactively contaminated place in the world."

As you can imagine, Chelyabinsk was not exactly a popular tourist attraction. Almost no one outside Russia had ever even heard of the place, and if they had, it was usually as a forbidding Cold War relic. But after Friday, February 15, 2013, Chelyabinsk was for a time one of the most famous cities on Earth.

Just after sunrise that morning, a twenty-yard-wide asteroid traveling at 42,000 miles an hour burned through the atmosphere east of the Ural Mountains in a trail of incandescent light and billowing dust and smoke. Moments later it exploded 14 miles up in the sky near Chelyabinsk, in a fireball that was brighter than the Sun. The bolide released 500 kilotons of energy—about thirty times greater than the Hiroshima atomic bomb—and sent a powerful shock wave radiating outward. The fireball transit and explosion were captured on scores of dashcams and phone cameras—and the reactions of the people filming the event were pretty much what you'd expect:

"Holy shit!" a man in one video exclaims in Russian. "What the fuck is that?" shouts another. "It's a bombardment! Get prepared!" another man warns. The videos show people staring in shock, dogs running in panic, cars swerving on icy streets, sirens and car alarms wailing.

"It was a light which never happens in life," one Russian blogger wrote of the bolide explosion. "It happens probably only in the end of the world."

It took about 90 seconds for the shock wave from the explosion to reach the city of Chelyabinsk, a time lapse that made its effects even worse. Alarmed by the brilliant fireball explosion seen in the distance, people indoors had run to their windows to see what was going on. That was the worst thing they could have done. They were still looking through windows when the shock wave hit and blasted the glass into little pieces, sending shards into faces, arms, eyes.

More than fifteen hundred people were injured, mostly by flying glass and being knocked to the ground. One woman had a broken back, and dozens of others who were closer to the fireball explosion suffered vision damage and sunburn-like ultraviolet burns. More than seven thousand buildings were damaged. The fireball explosion also sent thousands of pieces of the 12,000-ton asteroid falling as meteorites over a wide area southwest of the city, including one 1,200-pound meteorite that punched a 20-foot-diameter hole through the ice at nearby Lake Chebarkul. Amazingly, no one was killed by either the blast wave or the rain of stones.

Despite the destruction, as asteroid impacts go the Chelyabinsk event was relatively small. It was much smaller than the impact that formed Meteor Crater, and obviously infinitesimally smaller than the K-T extinction impact sixty-five million years ago. Still, it was the second-biggest known impact in the past century or so—smaller than the Tunguska Event in 1908 but larger than the previous second-place holder, the exploding asteroid that rained meteorites over the Sikhote-Alin Mountains in far-eastern Russia in 1947. And it was the only known asteroid impact in modern human history to cause mass casualties.

(The fact that all three of the most powerful known asteroid impacts in the past 110 years occurred in Russia doesn't mean that the asteroid gods have it in for Russians. It's just that Russia is a big target. Geographically it's the biggest nation on Earth, covering more than six million square miles and stretching out over eleven time zones. Statistically there's a 1-in-32 chance that any given Earth-impacting asteroid will hit Russia, as opposed to a 1-in-200,000 chance that it will hit tiny Luxembourg. Also, there could have been other, bigger impacts over the ocean or Antarctica or some other remote place before we had the technological means to detect them.)

Of course, the Chelyabinsk event produced front-page headlines around the world—"Surprise Attack: Meteor Explodes Over Russia," and "Shock Wave of Fireball Meteor Rattles Siberia" were a couple examples—and it led every TV news broadcast, complete with footage from all those dashcams and phone cameras. It was the most thoroughly covered asteroid-versus-Earth incident ever.

And predictably, the impact sparked countless wild rumors. One was that the asteroid had been "shot down" by a Russian military air defense unit. Another was that it was a giant satellite that crashed. And an ultranationalist Russian official named Vladimir Zhirinovsky—he's been known to wear a "Make Russia Great Again" hat—publicly blamed the strike on the U.S., declaring that "Those were not meteorites; it was Americans testing their new weapons." The impact also launched a "meteorite rush" as thousands of shards from the exploded asteroid were clawed out of the deep snow by local residents and then sold on the international meteorite market. Today a piece smaller than a sugar cube will set you back about fifty bucks. But be careful; there are a lot of fakes out there. Small meteorite pieces were also embedded in gold and silver commemorative medals that were handed out to some athletes at the Sochi Winter Olympics in 2014.

Meanwhile, the initial official reaction to the Chelyabinsk event among the astronomical community was—What the hell? Where did this thing come from? Here they were, with all their telescopes trained on 2012 DA_{14}, and then this cosmic cat burglar comes sneaking in through the back door, totally unexpected? It was a little embarrassing.

"Our emails just blew up," one NASA official recalls. "People wanted to know if DA$_{14}$ had strayed off course and we just missed it. It took everybody by surprise, and at first nobody really knew what was going on."

Astronomers pretty quickly got a handle on it, though. As it turned out, the Chelyabinsk asteroid had absolutely nothing to do with the close approach of 2012 DA$_{14}$, which was on a totally different orbit. It was just a coincidence—albeit an extremely spooky one.

As for the fact that no one had seen the asteroid coming, that was understandable. Even if the world's available telescopes had been searching for the Chelyabinsk asteroid, they almost certainly wouldn't have spotted it. It was just one more of countless small Near-Earth asteroids that hadn't been discovered, so no one even knew it existed, or where in space to look for it, or where it was going. Also, as it approached Earth the Chelyabinsk asteroid had the Sun at its back, which made it almost impossible to detect with ground-based telescopes; no asteroid-seeking telescope can point at the Sun without having its optics fried.

On the other hand, if like asteroid 2012 DA$_{14}$ the Chelyabinsk asteroid had been spotted a year earlier, if there had been sufficient resources in place to routinely detect such small, hard-to-spot asteroids, things would have been different. In that case the asteroid's orbit and point of impact could have been precisely determined, and emergency measures put in place before it injured hundreds of people at Chelyabinsk.

But it didn't happen that way. The Chelyabinsk impact came as a complete surprise, out of the blue. And it would be hard to imagine a more dramatic demonstration of the need for more and better Near-Earth Object detection programs, particularly those geared toward detecting smaller but still potentially disastrous ones. The Chelyabinsk event was almost universally described as a "wake-up call"—and this time the world actually woke up.

A few days after Chelyabinsk, the astronomers gathered at the United Nations NEO conference in Vienna formally called for creation of the "International Asteroid Warning Network" (IAWN) to promote and coordinate worldwide NEO detection and tracking programs. Officially sanctioned by the United Nations, IAWN members include NASA, the European Space Agency, the Chinese National Space Agency and others, all of them committed to sharing information and data and trying to get their respective governments to take the NEO threat seriously. The UN also sanctioned creation of the international "Space Mission Planning Advisory Group" (SMPAG, known as "SamePage"), composed of scientists and engineers who could provide technical advice on asteroid defense methods.

The dramatic videos coming out of Chelyabinsk also ultimately led to "Asteroid Day." In 2014 an international group of some two hundred scientists, former astronauts and others concerned about the NEO threat called for a dramatic increase in NEO detection programs, and for the creation of a global Asteroid Day to increase public awareness of the threat. The group included Apollo 13 astronaut Jim ("Houston, we have a problem") Lovell, Russian cosmonaut Alexei Leonov, Planetary Society CEO Bill "The Science Guy" Nye, British Astronomer Royal Lord Martin Rees, and British astrophysicist Dr. Brian May. May declared at a press conference that "The more we learn about asteroid impacts, the clearer it becomes that the human race has been living on borrowed time." (It may come as a surprise to learn that Dr. May, who has a PhD from Imperial College London, is also the lead guitarist for the rock band Queen. That's right. The man who gave the world "We Will Rock You" and "Fat Bottomed Girls" is helping to save the world from asteroids.) The group proposed June 30, the anniversary of the Tunguska Event, as the date for International Asteroid Day, a designation that was later formally endorsed by

the United Nations. Asteroid Day is now observed globally with
hundreds of events—lectures, forums, educational displays—live-
streamed around the world. That forum on asteroids I attended in
2017? That was an Asteroid Day event at the University of Arizona.

As for NASA, well, as tough as the asteroid impact was for the
people of Chelyabinsk, for the American space agency's Near-Earth
Objects program that little asteroid was a gift from heaven. As
one NASA official told me, "The joke was that we could use one
of those events every five years or so, just to keep people on their
toes—and to keep the funding coming." In the years after Chelyabinsk, NASA's NEO program saw its annual operating budget
double, and then nearly double again. (The 2019 fiscal year budget
for NEO programs is $150 million, another dramatic increase over
previous years—but as we'll see, it's still not enough to pay for
all of the NEO programs and missions we really need. As Casey
Dreier, director of space policy for the Planetary Society, points
out, it's only a tiny portion of NASA's $20 billion budget—less
than 1 percent. "The money we're talking about in terms of [NEO
detection programs] is minuscule," Dreier says.) Meanwhile, the
number of new NEO discoveries also more than doubled, from
about seven thousand in 2013 to nineteen thousand as of late 2018.

Clearly, almost two decades after America's space agency had
been somewhat reluctantly dragged into the NEO business, NASA
was now fully on board. And as if to highlight that fact, in early
2016 NASA introduced the American public to a phrase that
encapsulated the effort to protect Earth from hazardous space rocks.

It was called "planetary defense."

○

Lindley Johnson is NASA's first-ever Planetary Defense Officer. That's what it says on his business card—"Planetary Defense

Officer." He's the head of the space agency's Planetary Defense Coordination Office, which means he oversees every aspect of NASA's Near-Earth Object programs. Operating out of a third-floor office at NASA's headquarters in Washington, D.C.—there's a sign on the wall that says, "Every Day Is Asteroid Day"—Johnson is responsible for funding for the Catalina and other sky surveys, new asteroid detection technologies, and working with international NEO programs. In short, Lindley Johnson is America's asteroid-hunter-in-chief.

Not that he necessarily looks and acts the part of an aggressive space warrior; he doesn't walk around in combat boots and a formfitting Captain Kirk–style uniform. A soft-spoken, gray-haired man of sixty, Johnson favors slacks and a long-sleeved blue shirt with a small NASA logo on it—although he's been known to appear on video conferences wearing a T-shirt that says, "Asteroids Are Nature's Way of Asking: How's That Space Program Coming?" Low-key and utterly unflappable, with a wry, deadpan sense of humor, Johnson is the NASA official who will get the call if the boys at the Minor Planet Center and JPL discover an asteroid coming our way—and you get the feeling that if that happens, Johnson's heart rate won't so much as skip a beat.

"It's a big responsibility," Johnson says of his job. "But it's not what I would call particularly high-stress." Somehow that's kind of reassuring.

Johnson had long dreamed of working for NASA—but as an astronaut, not an administrator. As a kid he changed the spelling of his middle name from "Neal" to "Neil" because that was the way Neil Armstrong spelled it, and after graduating from the University of Kansas with a degree in astronomy and an ROTC commission in the Air Force, he applied for NASA's astronaut-engineer training program. Unfortunately the 1986 *Challenger* disaster put the program on hold, and Johnson had to move on. He spent most of his career in the Air Force Space Command and related fields, often working on applied space engineering projects of the highly classified sort. He also found time to earn a master's degree in engineering management from the University of Southern California.

Asteroids had barely even been mentioned when Johnson was an astronomy student at KU. But in the early 1990s, while he was assigned to the Air Force's Phillips Laboratory R & D center in New Mexico, Johnson met up with Tom Gehrels of the Spacewatch survey. Gehrels was trying to get the Air Force to help him with the new CCD technology for finding NEOs, and as a lot of people discovered when they worked with the legendary Gehrels, his enthusiasm for the subject was contagious. Johnson got hooked on asteroids.

"I thought this was something the Air Force should be involved in," Johnson tells me. "It was a significant threat, with national security implications."

It seemed logical to Johnson that the Air Force would play a role. After all, the Air Force was already deep into space surveillance, with telescopes and satellites and radar and other technologies deployed worldwide to detect missile launches and track satellites. So why not use some of those resources to track NEOs?

But it was a hard sell. For one thing, at the time Johnson was a young major, only a mid-level rank in the military officer corps, and he was going where no young Air Force major had ever dared go before. And of course there was the aforementioned "giggle factor," the idea that Near-Earth Object programs were a silly waste of time and money. You can almost see the senior brass laughing about it over drinks at the Officers' Club. Asteroids? *Asteroids?* This young major wants the United States Air Force to chase space rocks? He's gotta be kidding!

"There were all these colonels rolling their eyes and saying, 'That

Johnson, he's way out there on this thing,'" Johnson remembers. "The giggle factor was very much in play."

But as eager young majors will do, Johnson persisted. And the timing was propitious. The Cold War was ending, and the American public was expecting a "peace dividend" reduction in taxes for defense spending. Like all the U.S. military services, the Air Force was facing drastic reductions—and nobody was laughing about *that* at the O-Club. The Air Force had to look ahead—and space seemed like the logical place to look.

So the Air Force launched a study of future roles and missions it might fill in what it called the "transatmosphere" of air and space. The result was a 1994 report called "Spacecast 2020," which featured a number of "Global White Papers" on cool stuff the Air Force wanted to get into in the next quarter century. Subjects included "Defensive Counterspace," "Offensive Counterspace: Achieving Space Supremacy," and even "Counterforce Weather Control," which described potential methods of altering weather conditions for defensive or offensive purposes. Most of the white papers were classified, which was unfortunate; frankly, I would have liked to have known in 1994 what the Air Force was planning to do about the weather. But at the very end of the "Spacecast 2020" report there was an unclassified paper written by Major Lindley Johnson and a couple of his colleagues at the prestigious Air Force Command and Staff College. The report was titled "Preparing for Planetary Defense: Detection and Interception of Asteroids on Collision Course with Earth."

It was the first time that anyone had used the phrase "planetary defense" in connection with asteroids. Which means that not only did Johnson invent the phrase, he also invented his future title with NASA.

"I was trying to think of a term that would bring the Air Force and its defense role together with the astronomy community,"

Johnson says. "Not everybody in the science community liked it—they thought it was a little too militaristic—but it caught on from there, and now it's pretty generally accepted."

In the 34-page paper Johnson laid out the then-current state of Near-Earth Object knowledge as developed by Shoemaker, Gehrels, the Alvarezes, and others in the field: estimated NEO populations and sizes, impact energy yields, impact climate effects and so on. The paper discussed existing and potential technologies to address the NEO threat, urged the Air Force to establish a dedicated NEO command office and suggested spending a few tens of millions of dollars—in defense spending terms, a pittance—on the NEO threat. Finally the paper declared that, "While there is no reason to live in daily fear, there is a significant danger to our planet from an asteroid impact. . . . All that mankind lacks is a greater awareness of the threat, and the will to do something about it. Mankind must now prepare for planetary defense."

Once again, the timing was auspicious. Shortly after Johnson's paper was published, Comet Shoemaker-Levy 9 slammed into Jupiter—and the giggle factor was at least temporarily put on hold.

"Suddenly I was this visionary person," Johnson says, smiling. "Shoemaker-Levy made believers out of a lot of people, and the paper generated a lot of interest."

By the end of the 1990s the Air Force was involved in numerous aspects of NEO detection. As I mentioned earlier, the LINEAR survey at White Sands Missile Range was a joint Air Force, NASA, and MIT Lincoln Laboratory project. Eleanor Helin also used an Air Force Ground-Based Electro-Optical Deep Space Surveillance (GEODSS) telescope in Hawaii for her Near-Earth Asteroid Tracking survey from 1995 to 2007. In 1995 the Air Force also released previously classified data on hundreds of incoming small asteroid atmospheric airbursts that had been collected by spy satellites over the previous decades—although it did so a bit grudgingly, fearful that the data could be used to assess U.S. satellite capabilities and the weaknesses thereof. The Air Force continues to provide limited scientific access to that data today.

As for Johnson, while assigned to Space Command he continued pushing for an enhanced Air Force role in NEO programs—for his efforts, Glo Helin named asteroid 5905 Johnson after him, somewhat to the modest Air Force officer's chagrin—before he retired as a lieutenant colonel in 2003. Given his space background, the jump to NASA was a natural. He worked on a number of NASA projects, including the *Deep Impact* space probe to a comet—more on that later—and also oversaw the various components of NASA's Near-Earth Object program. It was kind of a lonely job; in the early days, Johnson was the only senior NASA manager assigned to NEOs. But Johnson kept advocating for a centralized and better-funded NEO program—which, as we've seen, paid off, with NASA's NEO budget almost doubling every year after 2012. In 2016 NASA finally brought its various NEO components under one bureaucratic roof, creating the Planetary Defense Coordination Office with Johnson as Planetary Defense Officer.

The function of the Planetary Defense Coordination Office really wasn't new; Johnson had essentially been doing the same thing for years. But the term "planetary defense" attracted a lot of attention—which was exactly what NASA intended. "Near-Earth Objects Program Director" was a perfectly fine job title, but "Planetary Defense Officer" was something a copy desk could get its teeth into. Newspapers and popular magazines couldn't get enough of it—"Meet America's Planetary Defense Officer!"—and Johnson spent a lot of time doing interviews. That's another function of the Planetary Defense Officer—getting the word out about potentially hazardous NEOs.

(The Planetary Defense Coordination Office [PDCO] should not be confused with NASA's Office of Planetary Protection

[OPP]. PDCO protects Earth against dangerous asteroids; OPP protects Earth against dangerous microbes that might be brought back by returning space vehicles, and also protects other planets and space bodies from contamination by dangerous cooties from us. It's why hardware bound for space bodies is constructed in clean rooms by guys in hazmat suits.)

So how does America's Planetary Defense Officer currently view the asteroid threat? Because it's Lindley Johnson, he views it calmly.

"I don't lie awake at night," Johnson says, noting that the statistical risk of a dangerous asteroid impact on any given day, or even any given decade, is pretty small; if he constantly thought that a big asteroid was about to hit Earth any minute he'd never get any sleep.

And in fact, the threat of an extinction-level asteroid impact seems a little less scary than it did when Johnson got into the NEO business a quarter century ago. Back then we had little or no idea how many of those big boys were out there, or what their orbits were. We didn't know we even *had* an NEO problem. Today, with an estimated 95 percent of the kilometer-plus NEOs identified and tracked, and with none of them posing an imminent threat, we can breathe a little easier.

Still, Johnson understands better than almost anybody that the long-term risk is there, particularly from the impact of smaller asteroids.

"Even an object a hundred meters wide would be a bad day for a city," he says. "Something larger than that, three or four hundred meters, would wipe out a statewide area, there's no doubt of that, and the effects would be felt nationwide. It could pose an existential national security threat. If one that size hit anywhere in the world there would also be global effects—not to the extent of a one-kilometer impact, where the atmosphere would be clouded with dust for the next five years, but it would be significant. It would cause climate change in a single afternoon."

True, it may not happen in our lifetimes, or in our children's lifetimes, or even in their children's lifetimes. But remember those immortal blind bumblebees buzzing around in the Superdome? Someday it will happen.

And what if it happens sooner than someday? If we discovered an asteroid heading our way, would there be anything that we as a nation, as a planet, as a species could do about it? Could we kill the thing or chase it away before it kills us, like they did in *Deep Impact* and *Armageddon*?

It's possible. There is, however, one rather large problem.

The problem is that when it comes to killing asteroids, it's not quite as simple as it looks in the movies.

CHAPTER 8

ASTEROID KILLERS

In the winter of 1967 a 21-year-old aeronautical engineering student named Joe Deichman spotted a somewhat unusual notice posted on a bulletin board at the Massachusetts Institute of Technology. It was a call to develop a space mission to save the world from an approaching asteroid named Icarus.

"Icarus must be stopped," the notice on the bulletin board declared, noting that if Icarus wasn't deflected or destroyed, the mile-wide asteroid's impact on Earth would release the energy of half a million megatons of TNT and likely trigger a new ice age. "No effort or funds will be spared in carrying out the detailed plan to be developed by the crack team of scientists and engineers assigned to this project. The major limitation is time. The program must use existent space technology and hardware, and it must succeed. . . . The problem solution may utilize a rocket to intercept the asteroid and nudge it from its course. Alternatively, it may be better to reduce it to rubble with a nuclear warhead."

Rockets! Space! Blowing stuff up! It all sounded pretty good to young Joe Deichman.

"This was at the height of the Apollo program, with the Saturn Five rockets and all that," Deichman, now a retired aerospace engineer, recalls almost a half century later. "Everybody was interested in that stuff, so it sounded kind of fun. We were going to blast the damn thing!"

Of course, the call to stop the hurtling asteroid—soon to be known as "Project Icarus"—wasn't real. It wasn't an emergency action program by NASA or the Department of Defense or any other government agency. Instead, it was the syllabus for Course 16.74 in Advanced Space Systems Engineering, a student research project overseen by a thin, slightly stooped MIT professor named Paul Sandorff. Twenty-one seniors and graduate students signed up for the course, Joe Deichman included, and for the next three months they worked on the problem. And what they came up with was historic, for two reasons:

It was the first time that anyone had specifically analyzed the technological requirements involved in preventing an asteroid strike on Earth. And it was almost certainly the first time that a student class project ever inspired a big-budget Hollywood disaster movie.

I'll get back to Project Icarus in a moment, but first some background on asteroid Icarus itself.

Asteroid 1566 Icarus was on a lot of people's minds in the mid-1960s. First discovered in 1949 by astronomer Walter Baade, and named after the mythical youth whose waxed feathers melted when he flew too close to the Sun, the mile-wide Icarus periodically comes close enough to Earth to rate as a potentially hazardous asteroid. It was predicted to come within four million miles in 1968, which was a comfortably wide margin—unless you believed the newspapers.

Two years before the Icarus Project began, an astronomer named Robert S. Richardson pointed out in a *Scientific American* article that there was a "remote possibility" that if Icarus's orbit was altered by gravitational forces by just a few degrees it could "some-day" collide with the Earth. Eventually those comments made it into newspaper headlines, but somehow the words "remote" and "someday" weren't included.

"Large Asteroid Is Headed for Earth" one American newspaper headline said. "Icarus Could Be Catastrophic in 1968" said another. Readers wrote in to ask if it was true that Icarus would hit Earth, or even if it missed whether its close pass would cause catastrophic tides and other widespread destruction. A theoretical physicist in Australia publicly suggested that the governments of the United States, Britain, and the Soviet Union were keeping a close eye on Icarus, and perhaps considering a mission to blow it up with nuclear weapons. In Britain a member of the House of Commons demanded to know if the government was deliberately keeping the impending collision with Icarus a state secret. A rumor out of Italy had it that the Vatican Observatory—yes, the Vatican has its own observatory—was busily plotting exactly where on Earth Icarus was going to hit.

Astronomers tried to knock down the doomsday rumors, repeatedly asserting that Icarus was going to be a four-million-mile miss. But some people remained unconvinced, as evidenced by this memorable headline in the *New York Times*: "Hippies Flee to Colorado as Icarus Nears Earth." The accompanying story noted that "several hundred hippies in flower-painted Volkswagen buses" had arrived in Boulder, led by a "holy man" named Don from Miami. Don predicted that when Icarus struck, "California will slide into the sea, there will be violent earthquakes and maybe even Atlantis will rise." Apparently only two places on Earth were safe from the asteroid, those being Boulder and Tibet, with Boulder being the easiest to get to in a flower-painted Volkswagen bus.

Meanwhile, back at MIT, Professor Sandorff noted the furor over Icarus and decided that it raised an interesting question: What exactly could Earthlings do if an asteroid was heading our way? Hence Course 16.74.

MIT in 1967 was not exactly a haven for either hippies or conspiracy theorists. So as you might expect, not everyone on campus took the notion of stopping an asteroid quite as seriously as Professor Sandorff did.

"There were a lot of jokes going around," Deichman recalls. Some students suggested building a giant trampoline to bounce the asteroid back into space. Others declared that all they had to do was move Earth out of the way. Even some of the students who signed up for the project were skeptical.

But they pretty quickly got into the spirit of the thing. For the purposes of the project the students assumed that Icarus would hit the Atlantic Ocean two thousand miles east of Florida on June 19, 1968, creating a tsunami that would send a 200-foot wall of water crashing into North America, including the greater Boston area—in which case, so long MIT! The impact would also send hundreds of millions of tons of climate-changing dust and water vapor into the atmosphere. It was like Professor Sandorff had said: Icarus must be stopped.

The students considered a number of options, including "soft landing" a space vehicle on the asteroid and firing rockets that would gently push the asteroid off course. But given the time constraints, the students decided the only way to deal with Icarus was to shatter or deflect it with nuclear weapons—lots of them. The twenty-one students in the class—all of them men; this was MIT in 1967—divided into seven groups, each assigned a specific part of the problem: orbits and trajectories, nuclear payloads, rocket boosters and propulsion, spacecraft design, guidance and control, communications, and economics and management. Experts in various fields, including a representative from NASA, were brought in for advice, and the group even traveled to Cape Kennedy in Florida to get a firsthand look at the giant, thirty-story-tall Saturn V rocket that theoretically would launch the nuclear warheads.

Eventually the students determined that stopping Icarus would require six Saturn Vs, each equipped with a 100-megaton nuclear warhead mounted in an Apollo spacecraft that would intercept the asteroid when it was millions of miles away. A succession of nuclear warhead detonations just above the surface of the asteroid would pack enough wallop to fragment Icarus or slightly alter its trajectory and cause it to miss Earth. The interceptors would also be accompanied by a monitoring satellite to measure the effects of the nuclear detonations and report back to Earth. This wasn't just vague theorizing, either; the students came up with detailed specifications for each piece of hardware and every aspect of the mission.

(There were rumors that the brainiacs on the Icarus team had devised a triggering device for the nuclear warheads that was actually superior to what the U.S. military had, at which point the Pentagon stepped in and classified that portion of the student project. If true, nobody's saying.)

Finally, at the end of the spring term, with the Icarus close pass still a year away, the Project Icarus students unveiled their plan at a presentation attended by a crowd of aerospace academics and some members of the press. The students gave their plan a 71 percent chance of preventing any damage to Earth from an impact, and an 86 percent chance of limiting the damage to the impacts of a few small fragments. But the asteroid-busting mission wasn't going to be cheap. The students estimated the total cost of stopping Icarus at $7.5 billion—about 1 percent of the U.S. gross national product—and acknowledged that it would cause some economic and political disruption, including delaying the Apollo Moon landings by about three years. But as the students put it, "In light of the consequences of a collision with an asteroid the size of Icarus, the possibility of such a collision, no matter how remote, cannot go unrecognized. The world must be prepared, at least with a plan of action, in case it should suddenly find itself threatened by what had so recently been considered a folly."

Predictably, the class project got an "A." But to the surprise of Professor Sandorff and his students, Project Icarus also got a lot of play in the press—and virtually all of it was favorable. *Time* magazine ran a story about it in its Science section—"Systems Engineering: Avoiding an Asteroid"—as did dozens of newspapers nationwide, with headlines like "Students Plot Attack on Asteroid" and "Bomb Asteroid Plan Proposed."

And as I mentioned earlier, Project Icarus also inspired a Hollywood movie. In 1979 American International Pictures released a $16-million-budget disaster film called *Meteor*, stocked with a host of "A-list" and "A-minus-list" stars: Sean Connery, Natalie Wood, Henry Fonda, Brian Keith, Martin Landau and others. The premise, as expressed on the movie posters, was simple enough: "It's five miles wide . . . It's coming at 30,000 m.p.h. . . . And there's no place on Earth to hide!"

The plot centered around efforts by American and Soviet scientists to blow the thing out of the sky with nuclear weapons, which they do in spectacular fashion, although not before New York City is destroyed by a rogue fragment. The movie was a disaster in more ways than one, grossing just $8 million in the U.S. and getting ripped by the critics, although it's developed something of a cult following in recent years. An announcement at the end credits MIT's Project Icarus with inspiring the film, although the producers weren't inspired to fork over any cash to the students involved. "Sure, they credited us," Joe Deichman says good-naturedly, "but they didn't pay us."

As for asteroid 1566 Icarus, as predicted it passed safely by Earth on June 14, 1968, at a distance of four million miles, leading the *New York Times* to announce on June 15 that "The world did not come to an end yesterday." Icarus continued on its merry orbits, and made its most recent close approach to Earth in 2015 at a distance of about five million miles; this time there were no reports of geriatric hippies with flower-painted walkers descending on Boulder or anywhere else. Icarus is due back for another five-million-mile flyby in 2043.

So could Project Icarus actually have worked? If in fact the asteroid had been on a collision course with Earth, could the MIT kids' plan have stopped it? Probably not. For one thing, the students envisioned using 100-megaton nuclear bombs, which didn't exist then and never have. The biggest nuclear bomb ever detonated was the Soviet Union's so-called Tsar Bomba, a 50-megaton monster so powerful that the Soviet air crew that dropped it near the Arctic Circle in a 1961 test was given only a fifty-fifty chance of survival. The biggest nuclear bomb in the U.S. arsenal in 1967 was about 25 megatons, but at the time the Pentagon was secretive about warhead yields, so the MIT students had no way of knowing that. Also, the students were hampered by an almost complete lack of knowledge about Icarus's shape and composition, both critical factors in any deflection or destruction attempt.

Still, Project Icarus was on the right track, and well ahead of its time. The students correctly concluded that there are basically two ways of avoiding an incoming asteroid:

We can nudge it off course.

Or we can blow it up.

○

Using nuclear weapons against an asteroid wasn't particularly controversial in 1967. As I said, Project Icarus got great press, and the MIT students' ingenuity was widely applauded. But attitudes about asteroid-killing nukes soon changed.

"In the nineteen fifties and sixties the idea of putting nuclear weapons in space seemed obvious, the inevitable next step," says Alex Wellerstein of Stevens Institute of Technology, an expert on the history of nuclear weapons. "But then people started thinking about space in a much different way."

That became clear in 1992, when NASA sponsored a scientific conference at the Los Alamos National Laboratory in New Mexico called the "Near-Earth Object Interception Workshop." It was one of the aforementioned series of conferences and workshops in the 1980s and early '90s tasked with assessing the NEO threat and deciding what to do about any impending asteroid impact. And it was a public relations debacle.

The workshop basically was divided into two factions, informally called the "lookers" and the "shooters." The lookers were those astronomers and other scientists who wanted to enhance NEO-detection programs, and if a large incoming asteroid was spotted, the world community could then decide what to do about it. The shooters, on the other hand, many of them associated with the defense industry, were those scientists and engineers who wanted to build anti-asteroid defense systems in advance of any specific NEO threat, including those posed by the more numerous small NEOs. Although non-nuclear asteroid-deflection methods were discussed—more on those later—most of the conference participants concluded that in most cases nuclear weapons would offer the only solution. One of the participants actually shouted "Nukes forever!" at the conference, although some insisted he was joking.

But a lot of people didn't think it was funny. With the U.S.-Soviet Cold War over, the workshop was widely viewed as an effort by elements of the "military-industrial complex" to find a new threat to justify continued funding for "Strategic Defense Initiative" military space technology—the much-derided "Star Wars" program. (The SDI program was proposed by President Ronald Reagan in the early 1980s to develop new weapons systems—space-based laser guns, for example—that could blast out of the sky incoming enemy nuclear-armed ballistic missiles.

The program was widely criticized for being destabilizing and technologically unfeasible, and after the Soviet Union's fall it was abandoned in the early 1990s.)

It didn't help that one of the shooters at the Los Alamos workshop was Edward Teller, the "father of the hydrogen bomb" and the man believed to have inspired the Dr. Strangelove character in the Stanley Kubrick film. To the anti-nuclear community Teller was the poster boy for the mad scientist mentality. Teller was already known for advocating some unusual uses for nuclear weapons, including using a hydrogen bomb to blast out an artificial harbor in Alaska and proposing the use of nuclear explosions to extract oil from tar sands. And now here he was at the interception workshop, calling for the development of a one-gigaton nuclear bomb to destroy large asteroids or comets—that's a billion tons worth of TNT, ten thousand times more powerful than any nuclear weapon ever conceived—and also suggesting a test-firing of a nuclear warhead into space to blow up an asteroid, just to see what would happen.

"He wanted to build a nuclear bomb big enough to blow Ceres!" recalls Clark R. Chapman, the renowned planetary scientist and the author (with astronomer David Morrison) of the influential 1989 book *Cosmic Catastrophes* that highlighted the asteroid impact threat. Ceres of course is the biggest asteroid in the Solar System, some six hundred miles wide.

The press and public had been barred from the Los Alamos workshop—because, the organizers said, they didn't want the discussions to be sensationalized by a bunch of over-excitable reporters. But the secrecy probably only made the situation worse. News about the workshop quickly leaked out—and the response was almost universally negative. A *New York Times* op-ed column by physicist Robert L. Park called the workshop "a revival meeting for Strategic Defense Initiative true believers [who] propose to defend Earth at tremendous cost against an imagined menace." A *Wall Street Journal* headline declared, "Never Mind the Peace Dividend, the Killer Asteroids Are Coming." A headline in the *San Jose Mercury News* magazine *West* put it this way: "Killer Asteroid Dooms Earth! (And if you believe that, Edward Teller and friends have billions of dollars worth of space weaponry to sell you.)"

"How can federal laboratories consider bombs to prevent a once-in-a-million-years catastrophe, when the ozone layer is deteriorating twice as fast as anyone predicted?" the *Mercury News* article demanded to know. "By what twisted priorities would a nation lead a charge against space invaders, while bucking world pressure to combat global warming?" It was the perfect meeting of "Ban the Bomb" and the with-all-the-problems-we-have-on-Earth syndrome.

"The idea of putting nuclear weapons in space is crazy," Steven Ostro, an asteroid researcher at JPL, told the *Washington Post* after the Los Alamos workshop. "I think that's insane. It could introduce new dangers that would dwarf that of asteroids." Other noted scientists, particularly the famous "science popularizer" Carl Sagan, joined in.

The opponents had a good point. No thinking person wants to militarize space with nuclear weapons; preventing that had been a stated goal of the world community ever since 1963, when the Partial Nuclear Test Ban Treaty prohibited nuclear bomb tests in the atmosphere or in space. That was supplemented by the 1967 Treaty on Principles Governing the Activities of States in the Exploration and Use of Outer Space, Including the Moon and Other Celestial Bodies—the Outer Space Treaty for short—which banned placing weapons of mass destruction in orbit, on the Moon, or anywhere else in space. More than a hundred countries are signatories to the treaty, including the United States.

And there was also the concern that space-capable nuclear weapons systems designed for defense against asteroids could be re-tasked for offensive action against somebody else. If, say, China or Russia—or even the U.S.—announced the development of a new space-directed nuclear weapons system designed strictly for planetary defense, could we take them at their word?

"It's a matter of who do you trust?" says nuclear weapons historian Wellerstein. "If somebody comes to you and says, 'I only want to use the weapons system for asteroids,' well, maybe so. But there are plenty of examples of things developed in one context being used in another context. And situations and governments change. You don't want a nice guy to build a defensive nuclear weapons system for asteroids that the tyrant who later takes over might use for another purpose."

Those are valid concerns. Still, even now, a half century after Project Icarus and a quarter century after the Los Alamos conference, when it comes to stopping an incoming asteroid there may be no practical alternative to nuclear weapons being used in space. Sure, there are a lot of non-nuclear technology ideas about how to stop an asteroid, and some of them are exceedingly cool in concept. But as practical real-world solutions they generally come up short.

Consider, for example, paintballs.

In 2012 the United Nations Space Generation Advisory Council sponsored a contest for young science professionals called the "Move an Asteroid Technical Paper Competition." The purpose was to come up with innovative new ways to address the asteroid threat. The winner was Sung Wook Paek, a graduate student in MIT's Department of Aeronautics and Astronautics—MIT again!—who suggested that the best way to deflect an asteroid is to send up a spacecraft to bombard it with white paintballs. The white paint would reflect more sunlight, with the photons bouncing off the white surface gradually nudging the asteroid

into a different trajectory that would cause it to miss Earth. Paek
used as a test case the 400-yard-wide asteroid Apophis—the same
Apophis that briefly made it to a "4" on the Torino risk scale in
2004. He figured it would take about five tons of paint and at
least a couple of decades to do the job on Apophis—which is one
of the big problems with the paintball idea. It would take a long
time, even for a relatively small incoming asteroid—time that we
probably wouldn't have. And the process wouldn't work on really
big or really small asteroids. Cool idea, but not very practical.

There are other so-called slow-push methods of asteroid deflection. In a variation on the paintball idea, some researchers have
suggested wrapping an asteroid in a kind of Saran Wrap film to
alter its absorption of heat from the Sun. As the asteroid rotates,
the dark side cools and heat radiates outward, giving the asteroid
a slight nudge—what's known as the Yarkovsky Effect. You could
also shoot an asteroid with lasers or a concentrated solar beam to
"ablate," or heat up, the surface, causing the heated material to
eject and produce a small thrust that would eventually alter its
orbit. You could use an "ion beam shepherd" spacecraft to bombard
it with high-speed ions and slowly push it off course. You could
use mass drivers, also known as electromagnetic catapults, to hurl
pieces of the asteroid out into space and create a small amount of
reverse thrust. Or maybe you could just mount a rocket engine
on the asteroid—an idea that was considered and rejected by the
Project Icarus students—and fire it up with enough thrust to
send the asteroid on a new course. The problem is that securing a
rocket to a small asteroid with almost zero gravity that's spinning
around at a high rate of rotation might be tough.

Still another potential solution for an asteroid destined to strike
Earth is the "gravity tractor." It's a simple concept. Just take the
biggest space vehicle you can get, park it near the asteroid and
let the space vehicle exert its own small amount of gravitational

pull on the space rock, eventually pulling it off course. Another version of this concept is the "enhanced gravity tractor," in which a spacecraft would briefly land on an asteroid, snatch up a boulder or other mass of material and then park itself and the boulder nearby. The boulder would increase the space vehicle's mass, and thus its gravitational pull, speeding up the deflection process. The beauty of the gravity tractor technique is that unlike most other deflection schemes, the composition of the asteroid wouldn't matter. Whether the asteroid was loosely packed rock and dust—known as a "rubble pile"—or denser rock or metal, the process could still work.

In fact, NASA actually had a plan to try out the enhanced gravity tractor idea. The $1 billion "Asteroid Redirect Mission" (ARM), which was scheduled for launch in 2020, would have snatched up a boulder the size of an SUV off an asteroid's surface and used its gravity to pull the asteroid slightly off course, after which the spacecraft would have hauled the boulder back home and deposited it into orbit around the Moon for further study. It was a bold and ambitious idea—too ambitious, as it turned out. The mission was canceled in 2017.

Each of the slow-push—or pull—deflection ideas has its own individual drawbacks, but they all share a common weakness: They take time, lots of time. To have any chance of success they'd have to be launched decades before an asteroid's predicted impact on Earth. So if the situation is a little more urgent, if an asteroid is due to hit us within years or months, there are two basic choices: Knock it or nuke it.

Knocking an incoming asteroid off its trajectory toward Earth would involve hitting it with hypervelocity "kinetic impactors." The concept is built around technology we mastered centuries ago—which is to say, the cannonball. Take the biggest, heaviest piece of metal you can put into space and crash it into the approaching asteroid at a speed of four or five miles per second. The idea isn't to shatter the asteroid, but just to slow it down a little bit—a tiny fraction of an inch per second. That may not sound like much of a speed reduction, but over years and millions of miles it adds up. So by the time the asteroid reaches the spot where it was going to hit Earth, Earth won't be there anymore. Our planet will have already sailed happily on in its orbit around the Sun and the asteroid will safely pass by. (You can accomplish the same thing by hitting the asteroid from behind and speeding it up just a little.)

Hitting a small object that's millions of miles away from Earth and traveling through space at tens of thousands of miles per hour isn't easy—but it has been done. In 2005 NASA's *Deep Impact* space probe launched an 800-pound chunk of copper and aluminum at periodic Comet Tempel 1; the impact blasted out a crater on the surface of the five-mile-wide comet nucleus that was 300 feet wide and 100 feet deep. The purpose of the mission wasn't to slow the comet down, but instead to study the comet's internal composition. Nevertheless, the impact did slow Tempel 1 down a little—very little, only a tiny fraction of a millimeter per second. But it illustrated the concept.

Of course, smacking a five-mile-wide comet with a kinetic impactor is one thing; hitting an asteroid just a few hundred yards wide is another. So NASA is planning another mission to test the kinetic impactor idea on a small asteroid. It's called the "Double Asteroid Redirection Test," or DART. Its target is asteroid 65803 Didymos, a half-mile-wide NEO that has its own little 150-yard-wide moon—informally known as "Didymoon"—that orbits around the asteroid just like our Moon orbits around Earth. The DART mission would whack the little moonlet with a 1,000-pound impactor traveling at about four miles per second, causing a slight change in its orbital speed. (In its current orbit, asteroid Didymos does not pose an imminent threat to Earth; DART would be a test, not an actual save-the-world mission.) The DART mission is scheduled for launch in 2020 or 2021, assuming the funding holds up, and it would be the first time that the kinetic impactor asteroid deflection method would be specifically tested in real-world—or real-space—conditions.

Kinetic impactors are just about everybody's favorite means of deflecting an incoming asteroid. They're a simple concept, they use currently available technology, and, best of all, they're non-nuclear. NASA calls them "the preferred method" of asteroid deflection. But they might not be effective against a loosely bound rubble pile asteroid; they might just punch right through it without slowing it down a whit. And NASA acknowledges that in many, perhaps most, incoming asteroid situations, kinetic impactors might take too long and probably wouldn't pack enough punch.

In 2018 researchers at Lawrence Livermore National Laboratory released the results of an asteroid deflection plan funded by NASA and the National Nuclear Security Administration. Titled "Options and Uncertainties in Planetary Defense," the study analyzed the capabilities of kinetic impactors, offered a case study of an asteroid-deflection mission and provided a detailed design for a future "planetary defense spacecraft."

The researchers chose as their test subject the asteroid 101955 Bennu, another leader in the perp walk of most-threatening Near-Earth Objects. Discovered in 1999 by the LINEAR sky survey, the 500-yard-wide Bennu makes relatively close passes by Earth every half-dozen years or so. It's expected to make a particularly close approach—within about 180,000 miles—in 2135. As of now astronomers are confident it won't hit our great-great-great-grandchildren on that pass. But if they turn out to be wrong, if Bennu does manage to plow into Earth, it will do so with the explosive force of more than a thousand megatons of TNT—not a global extinction-level event, but a very bad day for any country or region that happens to be in the way. For the purposes of the study the researchers posited that Bennu will indeed smack into Earth on September 25, 2135. (As I mentioned earlier, Bennu is also the target for NASA's $800-million-plus *OSIRIS-REx* mission, which is currently under way. In addition to collecting surface samples and gathering other scientific data, *OSIRIS-REx* will precisely measure how the asteroid's orbit is influenced by the Yarkovsky Effect, which as you'll recall is the slight nudge given an asteroid by the radiation of solar heat from the asteroid's surface. That's an important thing to know about asteroids in general, and it will help us determine if Bennu is going to come a lot closer to us in 2135 than we now think.)

The notional "planetary defense spacecraft" that would take on Bennu is called *HAMMER*, short for "Hypervelocity Asteroid Mitigation Mission for Emergency Response" and a leading candidate for the catchy-but-tortured acronym of the decade award. With a nod to concerns about using nuclear weapons in space, *HAMMER* was designed for a dual capability. It could either hit an asteroid with a 20,000-pound kinetic impactor "slug," or it could be armed with a nuclear weapon.

The 70-page "Options and Uncertainties" report is highly technical, full of complicated scientific calculations and diagrams. But its conclusion is simple enough. When it comes to stopping even a relatively small asteroid like Bennu, kinetic impactors aren't all they've been cracked up to be.

"Whenever practical, the kinetic impactor is the preferred approach," the study concluded. "But various factors, such as large uncertainties or short available response time, reduce the kinetic impactor's suitability and, ultimately, eliminate its sufficiency." The study found that even with a 25-year lead time, it would take more than a dozen *HAMMER* impactors to deflect an asteroid the size of Bennu away from Earth, and even then it might be too close for comfort, with the asteroid missing us by only five thousand miles or so. If we had only ten years' lead time, we'd have to launch as many as eighty kinetic-impact *HAMMER*s to deflect the thing, at crippling economic cost. And that's just for a relatively small asteroid like Bennu; anything bigger would be even more difficult. It would be like trying to slow down an oncoming 80,000-pound big rig by throwing marbles at it.

So what's left in the anti-asteroid arsenal? Like it or not, for now at least it's nuclear weapons.

There are two ways to use a nuclear detonation to stop an asteroid from hitting Earth. If there's enough time before the expected impact you can try to slow the asteroid down by detonating the nuclear device in a "stand-off" position tens or hundreds of yards away from the asteroid's surface. That would vaporize and throw out a thin layer of the asteroid's surface, providing a small amount of reverse thrust. In other words, you'd use the nuke to slow the asteroid down a little, not to break it up into little pieces.

The problem with that approach, though, is that like kinetic impactors and other deflection methods, using the stand-off nuclear method to deflect the asteroid would require lots of time. You'd have to do it years before the asteroid was due to strike Earth. Which brings us to the second nuclear option.

"There could be cases where the warning time is too short to deflect [the asteroid]," says Megan Bruck Syal, a co-author of the 2018 "Options and Uncertainties" study. "Then the path forward would be to vigorously break it up into many small pieces that are well-dispersed" on their own trajectories. The asteroid fragments would behave sort of like the pellets from a shotgun blast; the farther they traveled, the more they would spread out, and the less likely any one of them would be to hit the target—in this case, target Earth.

But either way you go about it, with a stand-off nuclear explosion or a shattering one, the idea of building and using nuclear weapons against asteroids still leaves some people feeling a little squeamish. For example, in 2013 *Mother Jones* magazine took note of NASA's interest in anti-asteroid nuclear weapons with a story headlined: "Blowing Up Asteroids: The Latest Excuse to Keep Nuclear Stockpiles?" The anti-nuclear website Nuclear-News.net ran a story with the headline, "More Jobs and Toys for the Nuclear Boys: Blowing Up Asteroids." The anti-nuclear sentiment isn't quite as strong as it was back in the 1990s, but it's there—enough so that NASA and others involved in devising nuclear-based anti-asteroid strategies tread carefully around the issue

"Folks sometimes have very intense feelings about whether they're comfortable with the [nuclear] approach," Syal says. She pauses for a moment, searching for the right words, and then adds—diplomatically—"Among the general public, those feelings may not always come from a full scientific understanding of what's involved." Syal explains that the detonations of anti-asteroid nuclear devices "would happen at astronomical distance scales away from the Earth, so there would be absolutely no risk of [radioactive] contamination. That's a question we get a lot."

"It certainly faces a hurdle in public perception," says Paul Chodas of the Jet Propulsion Lab's Center for Near-Earth Object Studies. "Personally, I think [using a nuclear device] would be a last-ditch resort. But I wonder if we would choose to use that method if we had not tested it ahead of time. And how are we going to explain testing a nuclear explosion in space? There are international laws in place [including the 1967 Outer Space Treaty] that would make testing very difficult, so it faces a hurdle there as well."

Actually, among the general public, or at least the American general public, there may not be as much resistance to using nukes against asteroids as people may think—perhaps because that's the way we've seen it done in the movies. In 2014 NASA surveyed about two hundred people, most of them non-scientists, at two forums it conducted on asteroids. The survey found that a majority—57 percent—supported using nuclear weapons against even small asteroids that had a good chance of hitting Earth, and 85 percent favored using them against larger and more dangerous asteroids. NASA called the results "surprising." Even more surprising, to me anyway, was that 5 percent of the respondents favored "no action" of any kind, even non-nuclear action, against a large asteroid that could wipe out North America. I guess some people really do look forward to the end of the world.

And how big an asteroid could the current technologies and methods handle?

"We can definitely handle anything up to a kilometer with current off-the-shelf technologies," Syal says. "Of course, nature could always come up with something bigger heading our way, but the orbits of most of the larger asteroids have been well characterized, so that's pretty unlikely."

And there's another big question involved in planetary defense against asteroids. Whether we use nukes or kinetic impactors or some other means of defense, at what point do we start? At what level of certainty of impact do we go to the effort and expense of actually building and deploying an anti-asteroid system?

As the immortal Yogi Berra probably never said, "It's hard to make predictions, especially about the future"—and that certainly applies to predicting asteroids. If they have enough data on a known asteroid, the guys at JPL can determine with a high degree of certainty if it's going to hit Earth anytime soon—within the next century or so. But with a newly discovered Earth-crossing asteroid, it takes time to get enough observations and data to predict an impact with absolute certainty.

For example, say the Catalina Sky Survey discovers a 300-yard-wide asteroid and the guys at JPL start crunching the numbers. At first they calculate the chances of the asteroid hitting Earth twenty years from now at 1-in-1 million. But as more follow-up observations come in, those odds change. Now there's a 1-in-500 chance of the asteroid hitting Earth in twenty years, and then the odds are narrowed down to a 1-in-100 chance. Sure, the scientists are concerned, and they're keeping a close eye on this asteroid. But what congressperson is going to vote to start spending billions of dollars on a space mission to deflect an asteroid that has a 99 percent chance of *not* hitting the Earth two decades from now? Imagine what the with-all-the-problems-we-have-on-Earth crowd would say about that.

So we wait. But for how long? Until there's a 5 percent chance that the asteroid will hit? A 10 percent chance? A 50 percent chance? The problem is that the longer we wait, the harder it's going to be to deflect the thing. And if we wait too long, deflecting the incoming asteroid could be impossible.

"That's the tricky part," says Lindley Johnson, the guy who would help advise senior NASA officials and Congress and the White House on what to do if we spotted an asteroid heading our way. "Most decision makers don't want to make a decision until it's absolutely necessary, and it's very possible to make a decision too late. So that's what's challenging. This isn't just about science; there are social, political, and economic factors involved as well."

And even if the political shot-callers finally decided it was time to do something about an approaching asteroid, building and launching an anti-asteroid spacecraft and actually getting it out to the asteroid to do its job would take years—years we might not have.

Aerospace engineer Bong Wie doesn't think we should wait for even another day. The way he sees it, we should stop talking and start building an actual anti-asteroid nuclear weapons system now, right this minute, and have it standing by in case we ever need it.

"We've been talking about this [planetary defense] for thirty years but nothing really ever gets done," Wie says. "We have the technology, but we need to engineer and assemble those technologies. Now it's time to do something."

Wie, an expert in spacecraft control and guidance systems, is the director of the Asteroid Deflection Research Center at Iowa State University. Born in South Korea, with a PhD from Stanford, Wie has received a total of $600,000 in NASA grants to design an anti-asteroid defense system. And as far as he's concerned, all of the other ideas about deflecting asteroids—gravity tractors, kinetic impactors, even the stand-off nuclear explosion approach—simply won't work. They just take too long.

"We aren't going to have ten or twenty years' warning of a definite impact," Wie says. "We can't even rely on five years' warning. We might not know with a hundred percent certainty that the incoming body will hit Earth until it's in its terminal orbit. The only option at that point is to explode the incoming body with a nuclear device. So we should have a system pre-deployed, waiting in a silo like an ICBM, that could be launched with only weeks or even days of warning time."

The system Wei envisions is a spacecraft he calls the "Hypervelocity Asteroid Intercept Vehicle," or HAIV—and it takes a page right out of *Armageddon*. You'll recall that in the movie Bruce Willis and his rag-tag crew of oil well drillers don't just slam a nuke into the asteroid's surface. Instead they drill down into the asteroid to plant a nuclear device below the surface, thus concentrating the nuclear bomb's explosive power. As a character in the film explained, it's like the difference between having a firecracker go off on your open flat palm or inside your clenched fist. The open palm explosion would burn you and hurt like hell, but it wouldn't do too much damage since most of the explosive force would be directed into the air. But if the firecracker explodes inside your clenched fist—well, you probably aren't ever going to open another ketchup bottle unassisted.

So Bruce Willis does just that, drilling a deep hole, planting a nuke and blowing the Texas-sized asteroid into little pieces, along with himself. Actually that was another of the reported 168 scientific impossibilities in *Armageddon*. There is no asteroid the size of Texas. As we've seen, the largest is Ceres, which is about six hundred miles in diameter, while it's more than eight hundred miles from Beaumont to El Paso. And even if there were such a monster asteroid, blowing it up into little pieces would require a nuclear bomb with a yield in the giga-gazillion megaton range.

Still, Professor Wie thinks Bruce was on the right track.

"The movie basically had the right idea," Wie says. "The overall effect of an underground explosion is twenty times larger than a surface explosion."

So the HAIV would be a two-step process. First the spacecraft would launch a hypervelocity impactor at the asteroid, gouging out a deep hole. A millisecond later the other part of the space-craft, armed with a nuclear weapon—depending on the size of the asteroid, maybe a megaton or so—would slam into the hole and detonate with what Professor Wie says would be dramatic results.

"The asteroid would be pulverized!" he says enthusiastically. "Into dust!"

True, a lot would depend on what the asteroid was made of, and how far away from Earth it would be exploded. A bigger asteroid exploded relatively close to Earth could result in fragments that would continue on their impact path. But Wie thinks most of those fragments would burn up harmlessly in the atmosphere—"Enjoy the meteor shower!" he says. And even if a few fragments did impact, that would be preferable to the alternative. As Wie puts it, "Would you rather have five hundred Chelyabinsk events or one five-thousand-times-the-Hiroshima-bomb event?"

As for any radioactivity in the fragments that burn up in the atmosphere, Wie and others say the radiation danger would be negligible—although others disagree.

And the cost of building and launching his HAIV system? Oh, about half a billion dollars, plus another half billion for a backup. Wie suggests it might also be advisable to test-fire the system at an asteroid.

Unfortunately, Wie says, there's not much enthusiasm out there for spending a billion dollars on a system to use nuclear weapons to blow up asteroids.

"Politically it's almost impossible to even talk about it," Wie says ruefully. "People don't want to accept the facts."

(In deference to the anti-nuclear sentiment, Wie and his team have also designed a kinetic impactor vehicle that could be used if the warning time for an incoming asteroid were short, just a matter of weeks or months. Called the "Multiple Kinetic-Energy Impactor Vehicle" (MKIV), Wei's spacecraft is not designed to deflect an asteroid off its course. Instead, it would bombard a small asteroid—150 yards wide or smaller—with multiple impactors like a blast from a shotgun, breaking it into small pieces no bigger than a yard across.)

Unfortunately, none of the asteroid defense systems discussed here—Lawrence Livermore's *HAMMER*, Bong Wie's HAIV and MKIV, among others—have actually been built, and there currently are no plans or funding to do so. They remain notional concepts.

So what happens if at some point in the near future—say, tomorrow—Lindley Johnson gets a call telling him that Kowalski up on top of Mount Lemmon has just discovered a new and significantly large asteroid, and that the Minor Planet Center and JPL guys have calculated it's going to impact Earth in three weeks. What do we do?

Actually, a month after the Chelyabinsk event in 2013, the head of NASA, Charles Bolden, was asked that very question. At a congressional hearing, members of the House Science, Space and Technology Committee asked Bolden what we could do if an asteroid of Chelyabinsk size or larger was discovered heading for New York City in a matter of weeks—and Bolden's reply left the attending congresspersons in stunned silence.

What Bolden said was, "If it's coming in three weeks—pray."

Well, I suppose a little praying can never hurt. But prayer isn't exactly a strategy. Fortunately, in recent years astronomers and engineers from around the world have been meeting regularly to address that issue. If an asteroid is heading our way in weeks or years or even decades, what should we do? What technologies are available, and how should we deploy them? What are the social, economic, and international effects of an impending asteroid strike?

In short, they're trying to come up with a *plan*. And to do that, they're using one of the same tools the U.S. military employs when planning for terrestrial threats to the nation.

War games.

CHAPTER 9

ASTEROID WARS

I t's May 2014, just over a year after an asteroid shook Chelyabinsk, and the folks at the U.S. Federal Emergency Management Agency (FEMA) have a little problem.

There's another one on the way.

It seems that the boys over at NASA have spotted this 300-yard-wide asteroid, called 2014 TTX, that's in an orbit that crosses Earth's path. Initially the Sentry impact-monitoring system at JPL gave this asteroid just the teensiest-weensiest little chance—about 1-in-1 million—of smacking into the Earth seven years from now, in the year 2021. But now, with more observations of the asteroid coming in, those odds are getting shorter. Over the next few weeks the odds go up to 1-in-500, then 1-in-100. It's still a pretty long shot, only a 1 percent chance of impact, but it's enough to put this asteroid on the FEMA radar. As the federal agency that will be responsible for pre- and post-impact disaster measures in the unlikely event of an impact, FEMA will have to keep an eye on this one—just in case.

But then the odds of an impact get better—or rather, worse. In August 2014 the impact odds are estimated at 6 percent—and it's time to start taking this thing seriously. Sure, there's still a 94 percent chance that it won't hit Earth. But again, this asteroid is about 300 yards wide, and if it hits Earth it's going to release about 700 megatons worth of energy—almost fifty times more than the one that blasted out a huge chunk of Siberian forest back in 1908. An air blast or ground impact of that magnitude could damage or destroy everything in an area of several hundred square miles. It wouldn't be an extinction-level event, but if it hit a populated area it would be a catastrophe unprecedented in recorded human history.

The White House and the congressional leadership have already been informed, and naturally the politicians' first instinct is to dither. In briefings with astronomers and NASA officials they demand to know how the egghead scientists can be so sure there's even a 6 percent chance that this damn thing is going to hit Earth seven long years from now. After all, they say, you boys have been wrong before. Remember that time back in 2004, when you said there was a 3 percent chance that an asteroid—what's its name? . . . Apophis?—was going to hit Earth in 2029 and got everybody all excited? And then later you came back and told us, Oops, there's really a zero chance that Apophis is going to hit. So why should we believe you now? The scientists try to explain the subtleties and uncertainties involved in calculating the orbital characteristics of newly discovered asteroids, but it's like trying to explain that just because it's been an unusually cold winter it doesn't mean there's no such thing as global warming. Some people don't want to listen.

Still, while the politicians aren't willing to spend significant money on the problem, they want more information. So now a host of federal and international agencies get together to figure out what to do. NASA, Department of Defense (DOD), State Department, the European Space Agency, the newly formed International Asteroid Warning Network (IAWN), and the international Space Missions Planning Advisory Group (SMPAG)—they all trade information on the asteroid deflection capabilities and methods of various nations. FEMA starts drawing up contingency plans for evacuations, delivery of emergency supplies, and so on—again, just in case. The impact, if it comes, is still more than six years away.

Meanwhile, though, the impact probability keeps going up. By February 2015, it's at 35 percent, and NASA has come up with a "risk corridor" projection of where on Earth it might impact. The risk corridor is a narrow line that stretches halfway around the globe, over Africa, the Atlantic, the American Southwest, California and into the Pacific. The asteroid could strike anywhere along that line, but the NASA scientists can't say exactly where. They don't have enough information yet.

Of course, there's been no way to keep this thing secret. Asteroid 2014 TTX was listed on the JPL Sentry risk table as soon as it was discovered, with the data on its orbit available to anyone with a computer; any backyard astronomer can download programs to compute the impact risks. Also, the follow-up observations required to further refine the asteroid's orbit are performed by dozens, even hundreds of professional and amateur astronomers around the globe—and when it comes to sharing information, astronomers are notorious blabbermouths.

So the story is out there, and the news media are going nuts. Sure, some of the coverage is responsible, with clear explanations about the risk level, possible deflection methods, and so on. But the tabloids are having a field day: KILLER ASTEROID TO HIT EARTH IN 2021! Social media and the Internet are even worse, with videos depicting an asteroid the size of Pluto smashing into the Earth and conspiracy theorists claiming it's all a government plot to—well, to do something nefarious. Harried NASA and FEMA public affairs spokespeople try to get the facts out, but it's an uphill climb. The public doesn't know what to believe.

But by early 2016 there's no doubt. NASA now puts the impact probability at 100 percent, and it has shortened the risk corridor.

According to their calculations, in September 2021, asteroid 2014 TTX is going to hit somewhere along a line over the Caribbean, the Gulf of Mexico, and the states of Texas, New Mexico, Arizona and California. It's time to knock this thing out of its orbit and save a good chunk of America from destruction.

But how? We can't launch existing ICBMs or anti-ballistic missile devices at it. They don't have the lift or the guidance systems necessary to hit an asteroid in deep space. Because of funding limitations, NASA hasn't been able to fully design—much less build and test—a spacecraft capable of delivering a deflecting payload to an asteroid. That DART mission designed to do a test-deflection on the tiny moon circling around asteroid Didymos? That's barely on the drawing boards by this time. To build the necessary spacecraft they'll basically have to start from scratch.

And what kind of payload will that spacecraft deliver? For months the planners at NASA and DoD and other agencies have been arguing over whether to use a nuclear warhead or kinetic impactors to deflect the asteroid. A nuclear detonation would offer the biggest deflection push, but it's controversial. According to a NASA survey, a majority of the American general public supports the use of nuclear weapons against the asteroid. But it's bitterly opposed by some scientists and most environmental groups, as well as much of the establishment press. Reporters want to know, What if the nuke blows up on the launch pad? Or in the atmosphere? What if the nuclear-weapon-bearing rocket goes haywire and smashes into New York City? The pro-nuclear-option scientists and engineers try to explain that none of these things will happen, but politically that option won't fly.

So the decision is to go kinetic. U.S. agencies—principally the Air Force—and the European Space Agency each design and build three kinetic impactor spacecraft, for a total of six. The hope is that at least two of them will hit the asteroid with enough force to deflect it. The effort costs billions of dollars, but it's a crash program, with the highest national priority.

Still, it takes two years to prepare the missions. It's made even more difficult by the fact that scientists really don't have a good idea of what the asteroid is made of, or even its precise size, both of which are critical factors in planning a deflection mission. There isn't time to design, build and launch a probe to examine 2014 TTX up close.

In August 2018 the six kinetic impactor spacecraft are launched, but three of them fail, either at launch or en route to the asteroid. The remaining three are believed to have hit the asteroid on March 1, 2019, but scientists can't say for sure. At that point in its orbit the asteroid is not visible from Earth, and again, there wasn't time to send up an observation space vehicle to monitor the effect of the kinetic impacts. We have to wait and see if the impactors diverted the asteroid.

Nine months later, in December 2019, asteroid 2014 TTX comes back into Earth's view and we have the answer. The kinetic impactors have deflected a major portion of the asteroid off its Earth-impacting orbit. But the impactors also knocked off a 50-yard-wide chunk of the asteroid—and that large fragment could possibly still hit Earth in 2021. We won't know for certain until the mini-asteroid gets closer and we can get more observations.

So again it's a waiting game. In the meantime, senior officials from NASA and other agencies are hauled into congressional hearings, where congresspeople demand to know why the mission failed, or at least partially failed. Billions of taxpayer dollars spent, and a piece of that damned thing is still coming at us? The NASA officials try in vain to explain the inherent technical difficulties in deflecting a small object moving at high speed when it's millions and millions of miles away, especially when you have a short time to design and launch the mission. They do not point out that if

Congress hadn't been so tight-fisted with funding in previous years, NASA could have had the capability to spot 2014 TTX further in advance, and could have had the deflection technology already in place to deal with the asteroid instead of depending on an untested program.

Now just about everybody wants to nuke this thing and be done with it. The White House orders a new crash program to destroy the asteroid, this time with nuclear weapons. Unfortunately, that isn't possible. We have plenty of nuclear weapons available, but there is no existing launch capability to deliver a nuclear weapon to the asteroid when it's still far away from Earth—and there's not enough time to build one. The Earth has shot its currently available deep-space-capable-rocket wad with the six kinetic impactor launches. If the asteroid fragment is heading for us, we'll just have to take the hit.

And the asteroid is definitely heading for us. In May 2021 NASA announces there is a 100 percent certainty that four months from now, just after noon on September 5, the 50-yard-wide asteroid fragment will impact somewhere along a 600-mile-long, 20-mile-wide risk corridor that begins in the Gulf of Mexico south of New Orleans, passes over Houston, Texas, and ends just northwest of Austin.

True, the impact of the fragment will not pack the destructive punch that the larger 2014 TTX would have had—but it's still going to be bad. The asteroid fragment will be traveling at a speed of 35,000 miles per hour and will release the energy equivalent of ten megatons of TNT—roughly the equivalent of a thousand Hiroshima bombs. If it hits the Gulf of Mexico it will send a tsunami up to ten feet high ten miles inland across Texas and Louisiana. If it hits the ground it will wipe out everything within a three-mile radius and cause major destruction several miles farther out. In the more likely event that it explodes in the air, the damage will be spread over an even wider area. The explosion and heat will damage or destroy buildings over an area of hundreds of square miles—and if they don't get out of the way, a lot of people are going to die.

In fact, even before it hits, the asteroid is already doing damage. As you might expect, housing and other property values plummet within the risk corridor. Who's going to buy a home or invest in a business in Austin or Houston when there's a chance that in a few months there won't even be an Austin or a Houston anymore? Businesses shut down, people are out of work. Wealthy people can afford to relocate, but most people can't. There's a flurry of phony "asteroid insurance" scams, and people are demanding to know "Who can I sue?"

There's more. The Texas Gulf Coast area accounts for a quarter of the U.S. oil refining capacity, and the Gulf itself is dotted with oil well platforms, so in expectation of the impact disruption, gas and oil prices rise dramatically. In response, the White House orders a freeze on gas prices nationwide, which causes an artificial gas shortage and long lines at the pumps. The Texas governor also freezes prices within the impact zone, but there's widespread price gouging and growing shortages of food, water, and other supplies. FEMA starts making evacuation plans and pre-positioning disaster relief supplies.

Then, a week before the impact, based on radar observations from the Goldstone and Arecibo observatories, NASA is able to draw a bull's-eye on the exact spot where the asteroid is going to hit. It's going to hit the city of Pasadena, a working-class suburb of Houston, home to some 150,000 people. The Texas governor has already activated the Texas National Guard, and working with FEMA he orders mandatory evacuations from a wide area around Pasadena, involving hundreds of thousands of people. But it's like evacuating in the face of a hurricane. Most people wait until the last minute, hoping the hurricane will veer off course and hit someplace else; despite what they've been told by public officials, they don't understand that asteroids don't work that way. So when they finally do leave, the roads and highways are jammed.

And as with hurricanes, some people flatly refuse to go. FEMA officially calls this the "public non-observance of instructions" problem. Hundreds, even thousands of people simply don't believe the government, or are unwilling to abandon their homes and property. Others see the asteroid as God's will, the beginning of the end of the world. Churches are filled up; so are the bars. Police and the National Guard can't force the hold-outs to go, because all emergency responders have to evacuate the impact zone as well.

Finally, on September 5, seven years after asteroid 2014 TTX was discovered, after all the dithering and debate and partially failed efforts to stop it, the 50-yard-wide space rock screams into the atmosphere in a fireball brighter than the Sun. And a few seconds later—goodbye, Pasadena.

Well, as you no doubt have already figured out, none of this happened. There is no asteroid 2014 TTX, and it's not going to hit Pasadena, Texas, in 2021. Instead, this was the actual scenario used by NASA and FEMA officials for a "Joint Asteroid Impact Tabletop Exercise"—an asteroid war game—held at FEMA headquarters in Washington, D.C., in 2014. I've added some details for dramatic effect, but the basic premises are exactly as they appeared in the tabletop exercise scenario: A 300-yard-wide asteroid, a seven-year warning period, the debate over nuclear versus kinetic deflection, the partially failed mission, the impact on Pasadena.

The purpose of the drill was to familiarize FEMA and other non-space-related agencies with the unique problems posed by an unlikely but still possible asteroid impact—something that most government officials, like most of us, had never spent a moment's time worrying about. Fifty government officials and scientists attended the exercise, representing a host of government and private agencies: State Department, Energy Department, Jet Propulsion Lab, the White House Office of Science and Technology Policy, the U.S. military Northern Command (NORTHCOM), European Space Agency, Lawrence Livermore and Sandia National Laboratories, the National Geospatial-Intelligence Agency, and even the U.S. Forest Service and the U.S. Department of Education.

And that wasn't the only such asteroid exercise NASA and FEMA have conducted. In 2013, not long after the Chelyabinsk impact, the two agencies held their first asteroid war game. That scenario postulated that a 50-yard-wide asteroid called 2013 TTX— the TTX stands for "tabletop exercise"—would impact near the mid-Atlantic seaboard with only one month's warning. There was no time for a deflection mission, so the asteroid hit the shallow coastal waters right on schedule, creating a tsunami up to 45-feet high in some areas. Another joint NASA/FEMA exercise in 2016 used a 100-yard-wide asteroid—2016 TTX—due to hit California in six years. Scientists and engineers considered a nuclear deflection mission—under anti-nuclear political pressure, the California governor said that was okay as long as it was launched from Cape Canaveral, not California—but as it turns out there wasn't time. After millions of people fled, the asteroid hit the Rose Bowl in Pasadena in 2022 with the force of 50 megatons.

(Wait a minute. First Pasadena, Texas, and then Pasadena, California? Home of JPL? "That's kind of an inside joke," says Paul Chodas of JPL's Center for Near-Earth Object Studies, who was the lead designer of the NASA/FEMA scenarios. "Pretty soon we're going to run out of Pasadenas.")

You may have noticed that all of the NASA/FEMA exercise scenarios used relatively small asteroids, in the 100- to 300-yard-wide range. There are a couple reasons for that. There are a lot more smaller asteroids out there than the larger ones, so the chances of an impact by a relatively small asteroid are correspondingly greater. As for larger asteroids, say a mile wide or bigger, using one of them would defeat the purpose of the exercise. If a miles-wide asteroid were about to hit the United States there would be no place to go. You'd have to evacuate the entire country—an obvious impossibility. And who knows? Maybe Canada and Mexico would build walls to keep *us* out.

Anyway, the idea of the federal government conducting asteroid impact drills isn't nearly as outlandish as it would have seemed twenty or thirty years ago. Chelyabinsk and the increasing number of asteroid close flybys featured in the news have gone a long way in making people at least dimly aware of the asteroid threat. Still, the "giggle factor" isn't completely dead. When it comes to asteroids and planetary defense, you can never completely get away from the wisecracks.

"You do get some of that, the people who joke about aliens or whatever," says Leviticus A. Lewis, a retired Navy officer and self-described "space geek" who is now chief of FEMA's National Response Coordination Branch. "So you have to explain what it's about. But most people are glad that FEMA and NASA are taking this seriously. We respond to all hazards, and this is something that could be a possible hazard. So we're not stepping out of our mission set."

As you might imagine, for government officials who hadn't spent a lot of time thinking about the asteroid threat—and most government officials would fall into that category—the exercises were a bit startling.

"It was an eye-opener," says California Governor's Office of Emergency Services senior official Dan Bout, who attended the 2016 exercise—the one that took out Pasadena, California. "The possibility of an asteroid strike hadn't even entered my mind before that. I had always assumed that there was a plan on the shelf somewhere, something like, 'Okay, we'll just nuke the thing.' But then the scientists explained that there really isn't that capability right now. That was surprising."

"It's a good opportunity to practice the 'whole of government' approach to planning," says NASA Planetary Defense Officer Lindley Johnson, who helped organize all three TTX exercises. "It gives emergency managers a chance to learn what would be involved in an impact event. And it gives those of us in the asteroid science community valuable feedback on what information is critical for their decision making."

Predictably, the NASA/FEMA exercises sparked a lot of press attention and talk on the Internet, with many sources calling them "doomsday drills"—and some suggesting that the exercises meant NASA knows that an impact is coming and they just don't want to tell us about it. As the RT television network put it, "FEMA NASA Doomsday Drills: Is Something Big Coming?"

After all, that's how it always happens in the movies. To cite just one example, in *Deep Impact* the U.S. government keeps a tight lid on the news that a world-destroying comet is heading our way, in the meantime secretly developing a deflection space mission and building vast underground bunkers to save a portion of the population from extinction if the mission fails. A full year goes by before aggressive young TV reporter Téa Leoni finally manages to get the story and forces the government to go public.

Wouldn't happen, couldn't happen, say people involved in the process.

"It'd be maybe thirty minutes before it got out," says Lewis, noting that astronomers routinely communicate new asteroid discoveries and positions with other astronomers around the world via the Internet. Lewis didn't say this, but to paraphrase an old joke, the three fastest ways to spread information about an approaching

asteroid are telephone, television, and tell-an-astronomer. When
it comes to their field of study, astronomers are eager to share—
and they wouldn't necessarily have to take orders from the U.S.
government.

"Observing asteroids is an international enterprise," JPL's
Chodas explains. "International observatories are involved, international private and government agencies are involved. There's no
way you could keep that information secret." In fact, both Lewis
and Chodas stress that an important function would be for government agencies to put accurate information *out*, not to try to keep
it in. "Effective communication with the public would be a major
issue to try to avoid the spread of crazy fake stories," Chodas says.
"We get enough of that with simple close approaches of asteroids."

(That's why every chart and graph used at the tabletop exercises,
all of which are publicly available on the Internet, has the word
"EXERCISE" stamped on it in big letters. The last thing they
want is for an official NASA chart of the predicted blast zone of
an asteroid hitting L.A. in 2022 to wind up as "evidence" on some
wacky Internet site or late-night conspiracy call-in radio show.)

So what's FEMA's plan for a possible small asteroid impact?
Actually, it isn't that much different than for any natural disaster.
Keep the public informed, prepare for evacuations if necessary, bring
in relief supplies, plan for post-disaster reconstruction. The only
difference is that there might be a long lead time to prepare, maybe
even years. No one has ever been able to predict an earthquake or
a hurricane years in advance, much less be able to do something to
prevent it from happening. But with an asteroid impact it's certainly
possible to predict the event years ahead of time—which actually
could complicate matters. As noted in the 2014 exercise scenario,
a long warning time would improve the chances of a successful
deflection effort, but it would likely create social and economic
problems even before the impact occurred. It's a unique problem.

And how would we react to the impending threat of an asteroid strike? Would we put aside our political differences and act for the common good? Would we as individuals resist fear and panic and behave in rational and decent ways? Would such a thing bring us together, or further drive us apart?

It's impossible to know the answers to those questions right now—although if we wait long enough, someday we're going to find out. But the human reaction to an impending or actual asteroid or comet impact has been the subject of countless fictional treatments, with the predictions ranging from hopeful to exceedingly gloomy.

For example, to bring up *Deep Impact* again, the film portrayed the U.S. government setting up a lottery system to determine who would be among the lucky eight hundred thousand Americans allowed into the massive and well-stocked underground bunkers to ride out the impending comet impact. But only people under age fifty were eligible for the lottery, and the government also "pre-selected" for salvation a couple hundred thousand supposedly critical individuals—political and military leaders, scientists and engineers, artists and writers, even Téa Leoni. Well, can you imagine how that plan would fly with the AARP? And how would the average American Joe react to the news that he's going to have to eat a comet impact while Téa Leoni goes on living? (Actually—spoiler alert—she doesn't go on living; she gives up her spot in the bunkers and is swept away by a giant tsunami.) Nevertheless, the film portrays most of the lottery losers as calmly accepting their fate, with a marked shortage of riots, murders, and other outrages to public order. Maybe so.

But another widely popular impact treatment, Larry Niven and Jerry Pournelle's 1977 novel *Lucifer's Hammer*, takes a much darker view. In their version, a worldwide barrage of fragments from a comet plunges California into a pre- and post-apocalyptic nightmare, complete with mass starvation, crazed religious zealots, murderous biker gangs and roving bands of human cannibals; there are even former pet goldfish that grow into monster carp after feeding on limitless supplies of human bodies. Like I said, it's pretty dark—although having lived in L.A. for twenty years I can't say it's all completely out of the question.

Of course, the government officials involved in the NASA/FEMA exercises weren't eager to speculate about post-impact cannibalism or corpse-eating giant goldfish. But they generally agree that given time to think about an impending impact, the public reactions are certain to be mixed—and that "public failure to follow instructions" will be a problem.

"Americans make up their own minds," Lewis says. "There are going to be some that want it to happen, there's going to be others that say, 'I don't believe you guys, I'm going to wait it out.' That happens whether it's a hurricane or tornado or whatever. This would be a different thing, but I think we can count on similar responses."

Again, the likelihood of a destructive asteroid impact in America in the next week or year or decade is statistically slim. The people at FEMA and other state and local emergency management agencies aren't tossing and turning all night worrying about an asteroid hitting us tomorrow. But as the tabletop exercises demonstrate, they're taking the asteroid threat seriously. And they're following the most basic rule of disaster preparedness: It's better to plan for an unlikely event ahead of time than to start planning when the unlikely becomes the certain.

○

For obvious reasons, all of the NASA/FEMA exercises used an impact in the United States for their scenarios; in such a case the

U.S. government would naturally take the lead in any deflection or disaster relief decisions. But it gets even more complicated when the international community is involved.

Consider, for example, the scenario used for an asteroid impact exercise at the 5th IAA Planetary Defense Conference, hosted by the International Academy of Astronautics in Tokyo in 2017. Some two hundred scientists and aeronautical engineers participated in that asteroid war game, some acting out the roles of world leaders, others meeting in groups to assess the technical problems and suggest courses of action. In previous Planetary Defense Conferences in Italy in 2015 and in Arizona in 2013, the impact exercises wiped out Bangladesh and dropped an asteroid in the Mediterranean Sea off the coast of France, respectively. This time, in 2017, it was target Tokyo. The scenario, also created largely by Paul Chodas of the Jet Propulsion Lab, basically went like this:

It's 2017, and astronomers have discovered an asteroid called 2017 PDC. The asteroid is some 250 yards wide, and the International Asteroid Warning Network has announced that there's a 1 percent chance it will impact Earth in 2027, ten years from now, during a future orbit around the Sun. We might be able to knock it off its collision course right now with just a couple of rockets equipped with kinetic impactor space vehicles, but nobody in the world has any rockets set aside for planetary defense, nor is there an impactor system standing by. And even if we were ready, the mission would probably cost about a billion dollars—a lot of money for a space rock that has a 99 percent chance of missing us. So no need to do anything now. We'll just keep an eye on it.

A year later, though, in 2018, things are looking grim. For months the asteroid has been unobservable as it traveled around the Sun. But now it's back in view, and based on new observations of the asteroid by NASA's Hubble Space Telescope and the National Astronomical Observatory of Japan, the probability of an impact in 2027 is now at 96 percent—a virtual certainty. And the narrow risk corridor along which the asteroid will hit has been determined to extend over the North Pacific, Japan, North and South Korea, and China. Tens of millions of people live in the risk corridor. Something has to be done.

At an emergency meeting of world leaders, the engineers and scientists from the UN-sponsored SMPAG recommend an immediate flyby mission to take a look at the asteroid and assess its exact size and composition. They also recommend two types of potential deflection missions, one using a fleet of kinetic impactor space vehicles, the other using two space vehicles armed with nuclear weapons. But there's widespread international opposition to sending nuclear weapons into space. Given its history and internal politics, Japan in particular strongly resists arming any space vehicles with nukes.

Immediately there's squabbling among world leaders as to who will supervise the deflection missions. Only the United States, Russia, China, India, Japan, and the European Space Agency have the capabilities to launch interplanetary space missions, and their national interests do not necessarily align. The U.S. is worried about sharing its technologies with China or Russia in any collaborative deflection effort, while China is hinting that it may launch its own deflection mission to gain prestige by single-handedly saving the world. Everybody is worried that the "nut case in North Korea"—Kim Jong-un—may think that the whole thing is an American plot against him and will launch missiles at South Korea, Japan, and Guam.

Who will bear the costs of the deflection missions is another issue. Almost everyone agrees this is a global crisis, but who's going to pay? Should the poorer nations get a free ride? Or should they somehow be required to contribute to a worldwide anti-asteroid fund? There's a small but growing backlash in the U.S. and some

European countries, led by nationalist politicians and media commentators demanding to know why any countries not in the risk corridor should have to pay the escalating costs. Their slogan is "Not Our Problem!"

And there's another complication. If the mission planners use kinetic impactors against the asteroid, each strike will move the asteroid's impact zone on Earth a few hundred miles east or west along the risk corridor. If enough of the impactors hit the asteroid it will miss Earth entirely. But if some of the kinetic impactor spacecraft fail, and the asteroid doesn't get enough of a push, it will simply change the spot along the risk corridor where the asteroid will hit.

So do you try to nudge the asteroid in a way that will move the risk corridor eastward or westward? For technical reasons it would be easier to move it westward, so if the asteroid does hit Earth it will strike somewhere in the remote Gobi Desert in Mongolia—but as you might expect, that idea doesn't play well in Ulaanbaatar. The planners ultimately decide that if the asteroid can't be deflected entirely away from Earth, they'll at least deflect it enough to send it into the deep waters of the Pacific east of Japan. There is some danger of a small tsunami, but everyone agrees that would be preferable to a land impact; at one meeting a spacecraft engineer jokes, "We'll just hand out bathing suits."

In 2020 the flyby mission to the asteroid is completed, providing valuable information on its exact size and composition and further refining its trajectory. It's now certain that unless it's deflected or destroyed, the asteroid will strike Tokyo in 2027. The asteroid will hit with the force of at least 100 megatons, blasting out a two-mile-wide crater and knocking down buildings for thirty miles in every direction. Unless there are mass evacuations, twelve million people will be killed or injured.

Faced with an existential threat, Japan finally agrees to the use of nuclear weapons against the asteroid—but only as a last resort.

The U.S. and the European Space Agency design and launch two spacecraft armed with nuclear devices, while Russia builds another nuclear-armed spacecraft as a backup; those nations comprise what's known as the "Plan B" group. The U.S. and ESA missions will rendezvous with the asteroid for a stand-off detonation to slow it down by a tiny fraction—but they'll do so only if the kinetic impactors fail. Meanwhile, a fleet of five kinetic impactor space-craft is already heading for the asteroid; they're timed to strike the asteroid in 2024, four years from now.

In the meantime, Japan is suffering, economically and psycho-logically. Life in Tokyo is fraught with fear and uncertainty, and the world's fourth-largest economy is in shambles, with worldwide repercussions. As in Texas in 2021, property values in the Tokyo area plummet, forcing a wave of bankruptcies as banks have to call in their loans to meet capitalization requirements. Industry is fleeing the impact zone, and obviously no new investments are coming in. Public support is growing to use the nuclear weapons immediately and not wait for the impactor missions to arrive.

Finally, concerned that the nuclear-weapon-armed space vehi-cles will malfunction if they wait too long, the "Plan B" nations decide to act on their own. In 2024 they detonate one nuclear device a kilometer away from the asteroid, giving it enough push to make it miss Earth entirely—but just barely. Three years later the asteroid misses Earth by a mere 600 miles, and the world goes "Hooray!"—or, in Japanese, "Yatta!" The downside is that the Earth's gravitational pull has altered the asteroid's orbit, and now there's a chance that it could still come back and hit us in the future.

Well, again I've added a few touches, such as the "Not Our Problem!" slogan; on the other hand, the pass-out-bathing-suits-for-a-tsunami crack was real. I also omitted some technical details, including the fact that the threatening asteroid turned out to be a binary—that is, a big asteroid being orbited by a smaller one, a complication that had the mission planners tearing their hair out. And I should point out that in deference to the nukes/no-nukes debate, at the Tokyo conference there actually were two alternative outcomes decided upon: the nuclear deflection that I described, and also a successful kinetic impact mission.

But the rest is pretty much as the war game played out. And even though it was just an exercise, passions among the participants often ran high—especially on the nuclear issue. For example, when the mission-planning group voted to recommend using the nuclear weapon first, without waiting to see if the kinetic impactors would work, one group member called out in seemingly genuine anger, "But we promised the world that would be a last resort!"

"A bit of that was improvisational theater," says planetary scientist Clark R. Chapman, who has attended almost every planetary defense conference and workshop for the past quarter century, including the 2017 conference. "But people do tend to be passionate about things they're deeply involved in, and they do have very strong views."

"There are a lot of difficult issues involved," says Paul Chodas of JPL, the chief designer of the conference's asteroid war-game scenario. "Technical, political, economic, moral, legal. What are the international legal issues involved in using nuclear weapons in space? What are the legal and moral issues involved in deflecting an asteroid from one part of the Earth to another? These are the kinds of things we're trying to figure out."

The notion that we might accidentally or intentionally nudge an asteroid to miss Point A on Earth and hit Point B instead is a particularly thorny problem. And it raises an interesting question. What if sometime in the future some nefarious entity decides to weaponize asteroid deflection technology and methods? In other words, what if instead of deflecting an asteroid to miss Earth, someone deflects it to hit a very specific location on Earth? Would such a thing be possible?

Well, almost certainly not with today's technology and level of asteroid knowledge. Nudging an asteroid to hit a broad ocean or a vast empty desert would be tough enough; fine-tuning it to target a city would be even harder. You'd have to find just the right asteroid, on just the right orbit, and you'd have to give it just the right amount of push; a bit too much or a bit too little could throw it hundreds of miles off its planned target. It would be incredibly risky, with a distinct possibility of a serious backfire. For example, say that France decided to take out Berlin by deflecting an asteroid. The tiniest miscalculation in adjusting the speed of the asteroid millions of miles away from Earth could mean the asteroid would miss Berlin and hit somewhere else—say, 546 miles southwest of Berlin, in which case we wouldn't always have Paris. Besides, we already have plenty of perfectly efficient tried-and-true ways to inflict mass murder on one another. Why use asteroids?

So for now at least, asteroids-as-weapons seems like science fiction. (Actually it is science fiction. You may recall that in the 1997 film *Starship Troopers* the insect aliens direct an asteroid to destroy Buenos Aires; there are numerous other fictional accounts of weaponized asteroids.) But who knows? As asteroid deflection technologies and techniques become even more sophisticated, maybe in some future time one clique of human spacefarers will be able to direct small asteroids to take out another clique's Moon bases or Mars colonies or asteroid-mining operations. If that ever happens we will have come depressingly full circle from where mankind was a hundred thousand years ago; we'll be waging war by hurling rocks at each other.

There were some other issues raised at the 2017 Planetary Defense Conference—such as comets. You may have noticed that all the scenarios used at the various planetary conferences employed asteroids, not comets, as Earth's nemesis. That's because there are a lot more Near-Earth asteroids out there than comets, so the risk of a comet impact is pretty low. Which is a good thing. An incoming comet probably wouldn't be spotted until about a year before a possible impact, which wouldn't give us time to do much about it—or even to determine if an impact was certain. Also, comet nuclei are generally bigger than Near-Earth asteroids, and they travel much faster—more than 125,000 miles per hour or so. Given all that, a kinetic-impactor deflection attempt probably wouldn't work against such an object; as one conference participant put it, "Kinetic impactors wouldn't stand a chance in hell against a comet!" We might be able to knock out a small comet with a nuclear weapon if we had a nuclear-armed space vehicle standing by ready for launch—which currently we don't. So as it now stands, it's likely that all we could do in the face of an impending comet impact would be to cross our fingers—and finger crossing doesn't make for a very good war game.

Another issue at both the 2017 conference and the NASA/ FEMA exercises was appropriate terminology, especially concerning nuclear weapons. Participants were repeatedly urged not to use words like "nukes" or "nuclear weapons," but to say "atomic deflection devices," or better yet, "ADDs." Officials who might be called upon to explain the effects of cosmic impacts to the public were also advised not to express them in military terms, as in "ten megatons," or "the equivalent of ten million tons of TNT"—and for God's sake, never say something like "the equivalent of a thousand Hiroshima bombs." It may not sound like a major issue, but it's the sort of thing that people in the asteroid impact planning business have to think about.

In any event, as with most war games the purpose of both the NASA/FEMA tabletop exercises and the IAA Planetary Defense Conferences impact scenarios wasn't to determine which side would win—Earth or asteroid?—but to raise questions. What are our current capabilities? What are our limitations? What technologies and resources could be realistically developed to meet an asteroid impact threat in three or five or ten years? What social, political, and economic issues would we face? The point is that if we ask ourselves those questions now they'll be easier to address if an asteroid is ever discovered on a collision course with our planet.

And it's a continuing process. The next IAA Planetary Defense Conference is scheduled for the Washington, D.C., area in 2019, and the debates will go on. Paul Chodas of JPL, who's designing the impact scenario for the upcoming conference, wouldn't tell me which city he's planning to clobber with an asteroid at the Washington conference. But given the simulated havoc he has already wrought on Pasadena, Texas, and Pasadena, California, in earlier impact scenarios, I'm guessing that the nearby city of Pasadena, Maryland (population 25,000), is a likely candidate for ground zero.

〇

So where do we currently stand in terms of planetary defense against Earth-impacting asteroids? Are we ready and able to defend our planet against an impending asteroid threat?

The answer, in 2018, is yes, no, and maybe.

It's true that if we have enough warning—again, time is the most critical factor—we have the *theoretical* ability to deflect or destroy an asteroid up to about half a mile wide. (That anything bigger than that will threaten Earth anytime soon is highly unlikely—at least that's the consensus among astronomers.) We have the technology on hand or in development to launch an asteroid defense mission—perhaps kinetic but more likely nuclear—with at least a fair chance of success. The problem is that we currently lack an actual spacecraft to perform an asteroid deflection or destruction mission—and, apparently, we also lack the will to build one. Again, the *HAMMER* and HAIV and other asteroid destruction/ deflection systems described earlier are conceptual designs only; there are no current plans or funding to actually build them, or any other dedicated planetary defense space system.

Of course we should continue to explore asteroid mitigation technologies, as we're planning to do with the kinetic impactor Double Asteroid Redirection Test that NASA has in the works. But if we really want to get serious about planetary defense, we should start building an operational anti-asteroid space system like *HAMMER* and HAIV now and save it in storage for a rainy day—that is, the day that an asteroid or comet threatens to rain down on Earth. That might be a tough sell politically and economically, especially in the absence of any imminent threat. But if we wait until the eleventh hour to start preparing, we could get hit by an asteroid at 11:01.

Also, and perhaps most critically, we lack sufficient intelligence on the enemy. As we've seen, we've already identified the vast majority of large Near-Earth Objects, and have been able to predict their movements far into the future—with the happy result that none of them are likely to collide with Earth for at least a century or longer. But there are still thousands upon thousands of potentially hazardous smaller asteroids that haven't been discovered—and any one of those could be on a collision course with Earth in weeks or months or years. We just don't know. As has often been said, we need to find them before one of them finds us.

So we have to beef up our intelligence capabilities—and not just through continued funding for Earth-based asteroid detection programs like Catalina Sky Survey and Pan-STARRS. Launching the space-based NEOCam infrared telescope would provide intelligence data not only on the location of previously undiscovered Near-Earth Objects but also on their exact sizes and compositions—critical information for any asteroid defense plan. NEOCam's half-a-billion-dollar cost would be more than offset if the satellite discovered an imminent Earth-threatening NEO—and we'd be happy we spent the money.

NASA's *OSIRIS-REx* billion-dollar mission to Near-Earth asteroid Bennu will also help us gain valuable intelligence on its characteristics—but Bennu is just one NEO out of many. We need to gather intelligence on as many NEOs as we can, and private companies like Deep Space Industries and Planetary Resources could be a huge help with that. By sending out their fleets of small robotic space vehicles to explore potentially lucrative asteroids they will also gather intelligence that is critical to planetary defense— and in mining asteroids the companies will also develop techniques useful for future asteroid deflection. Cooperation between those private companies and government agencies in developing new space technologies won't just benefit the companies' stockholders. In the long run, it could benefit everyone on Earth.

And finally, to defend our planet from Near-Earth Objects we have to convince the national and international public that Earth actually needs defending. The Chelyabinsk impact and news coverage of asteroid close flybys have helped to do that; so have events like Asteroid Day. But human memories are often short, and the average person has plenty of other things to be concerned about. Without over-selling it, we need to stress that while an asteroid impact is a low-probability threat in any given lifetime, it's also an enormously high-consequence event, one that could be catastrophic not only for the immediate impact zone but also for the entire world's climate and economy. It's been said that if the dinosaurs had had a planetary defense program they'd still be here. Well, we humans do have a planetary defense program, or at least the beginnings of one. But it will require widespread public support for it to be effective. We can't let the with-all-the-problems-we-have-on-Earth attitude carry the day.

If we did all these things, we could be ready for an impending asteroid strike—maybe. There's always the chance that there's an asteroid or comet out there that would be too big and too bad for us to handle with any current technologies. We can't prepare for everything—but that doesn't mean we shouldn't prepare for anything.

The good news on the Near-Earth Object front is that governments around the world have finally recognized the impact threat as a legitimate national and global security issue—something that would have been almost unimaginable thirty or forty years ago. For example, in June 2018 the White House and the National Science & Technology Council released a "National Near-Earth Object Preparedness Strategy and Action Plan," an 18-page summary of what's being done to address the NEO issue—and what needs to be done in the future. Drafted by a multi-agency working group, the report recommends many of the steps mentioned above: enhancing NEO detection programs, including detection of smaller asteroids— 50 to 150 yards wide—that could pose local or regional threats; increasing international cooperation through agencies such as the International Asteroid Warning Network; conducting more impact exercises involving FEMA, NASA and numerous other government agencies; and continuing to develop technologies and strategies for asteroid deflection missions. Critically, the report stresses that the deflection strategies should include conducting a series of actual "flight demonstrations" by spacecraft to test various deflection methods such as gravity tractors, kinetic impactors, and nuclear devices—although in the case of nuclear devices, any nuclear detonations in space would be simulated. The bad news in the "Strategy and Action Plan" is that while it recommends that these steps be taken, it can't mandate them or provide funding. The money to accomplish those goals will have to come from our political leaders, and by extension the general public—which is to say, you and me.

So there's cause for both optimism and pessimism. Bruce Betts, chief scientist for the Planetary Society and a keen observer of NEO detection and mitigation programs over the past twenty years, puts it this way: "We've made a lot of progress in the past two or three decades. We know where the larger [NEOs] are, which is good, but we're still woefully ignorant in terms of the million or so other smaller objects that could cause city-sized disasters. And we're still novices in asteroid deflection techniques. We've accomplished a lot, but we're not yet at the point where we can effectively protect Earth. The big question is, will the progress we've made continue? Let's just say that I'm cautiously optimistic."

Again, the chances that we'll be confronted by a potentially world-changing asteroid impact in our own lifetimes are pretty slim. As the scientists like to say, the risk is small but non-zero. But as with any non-zero risk, given enough time, it will happen.

It's not a question of if.

It's only a question of when.

And at this point, we can only hope that the world will be ready.

EPILOGUE

It's a warm and typically windblown evening in the Arizona high desert, and the Sun is setting on Meteor Crater. Shadow is creeping across the crater floor, and the crater walls are alight with sunset colors, the colors of fire. The last visitors are piling into their SUVs and minivans, and the crater staff members are getting ready to lock up. I have time for one last look.

I've come back here to remind me what this book has been about. It's one thing to read about the enormous power of cosmic impacts, to calculate the energy and measure the radiating circles of destruction. But here you can actually *feel* it. Here you can feel the almost inconceivable violence of that moment long ago, when the heavens rained down in fire and fury and left this massive scar on the face of Earth. It's a physical sensation of wonder, and awe.

And it inevitably raises the question: What if such a thing were to happen now? What if our little asteroid of long ago were to make its fatal rendezvous with Earth at this exact spot fifty thousand years later than it actually did—say, tomorrow? What would be the consequences?

Actually, if you had to choose a place on Earth's land surfaces for the asteroid to impact, this barren, sparsely populated swath of Arizona desert might be as good as any. True, it would cause damage. After the impact, blast winds in excess of 200 miles per hour would sweep across Interstate 40, bowling over 18-wheelers like mastodons and sending minivans tumbling like Yesterday's camels; survivors crawling out of the wreckage would be scoured by a rain of rocks and metallic pellets, like iron hail. The Bar T Bar Ranch would be no more, and the Mobil station at Exit 233 would have sold its last gallon of Super Plus. Tourists at the Standin' on the Corner park in Winslow would feel a tremor rising up from the ground, and would be treated to the panic-inducing sight of an atom-bomb-like mushroom cloud rising into the atmosphere. If our asteroid hit here today there would be deaths, each one a tragedy, but the toll would be relatively small. The world would go on its way.

But imagine instead that our little asteroid collides with Earth 2,000 miles farther east. Suppose that it arrives without warning—which, given our current ability to detect small incoming asteroids, it certainly could—and that it strikes New York City or Washington, D.C. Superimpose this mile-wide crater and its ten-mile-wide circle of total destruction over Manhattan and try to imagine Times Square atomized and a blast wave of nuclear bomb proportions radiating out through Jersey City, Hoboken, Queens, the Bronx. Make the Washington Monument ground zero and imagine our nation's capital and everyone in it vanished in the briefest instant. Put this crater and the asteroid that created it into any heavily populated portion of our nation or the world and try to grasp the magnitude of the destruction. Millions would die, millions more would suffer. It would be the single greatest natural catastrophe in the history of our species.

And as you ponder that, recall that the asteroid that made this massive crater in the Arizona desert wasn't anything special by asteroid standards. Compared to an Apophis or a Bennu, it was almost insignificant—and yet it still possessed so much power that under other circumstances it could have altered the history of the world. Those are the kinds of thoughts and visions you have when you look at Meteor Crater.

And Meteor Crater is also a reminder of just how far we've come in our understanding of the wonders and dangers that lurk in space.

A little more than two hundred years ago, before the Celestial Police began their search for a missing planet, no one knew that asteroids like the one that made this crater even existed. Now we know that asteroids exist in the millions, and we're sending out spacecraft like *OSIRIS-REx* and others to explore their strange and mysterious world.

A hundred years ago, when Daniel Barringer was feverishly searching this crater for an asteroid fortune, the scientific establishment ignored him or laughed at him. Now NASA and companies like Deep Space Industries and Planetary Resources are developing technologies to harvest the staggering riches that asteroids contain, riches that might one day save a depleted Earth.

Half a century ago, when this crater helped inspire Gene Shoemaker and a few others to begin sounding the alarm about Near-Earth Objects, people still giggled and rolled their eyes, and many scientists scorned the notion that a small space body could inflict a worldwide catastrophe. Now scores of astronomers like Richard Kowalski are on their mountaintops, searching the skies for potential dangers and building up databases that one day could help save our planet from destruction—the kind of destruction we watched firsthand when Comet Shoemaker-Levy 9 slammed into Jupiter.

And even just twenty years ago, few would have guessed that an asteroid much smaller than the one that made this crater would come out of nowhere and rattle a Russian city, shocking the world and leaving hundreds of people bloody and screaming in fear. Now

Lindley Johnson and hundreds of other scientists and engineers and emergency planners are working diligently to figure out how to stop an incoming asteroid, and how to prepare for it if it can't be stopped.

Yes, we've come a long way. But we still have a long way to go.

As I mentioned in the introduction, when I began this book I really didn't know much about asteroids and impacts on Earth—and compared with the brilliant people in the field that I've talked to, I still don't know much about them. I've learned a lot of cool facts—and a few alarming ones—about cosmic impacts, but I remain a non-scientist, with a non-scientist's limited view. For example, I'm still trying to get my head around the finer points of the Yarkovsky Effect.

But I do know this much. The asteroid threat is real. It's not some cosmic boogeyman cooked up by astronomers and aerospace engineers in search of funding, or merely fuel for conspiracy theorists who see aliens throwing rocks at us. The asteroid threat is science fact, not science fiction. And it needs to be taken seriously.

True, the threat is not so imminent that you and I need to think about it every day—although we should be glad that there are people out there who do. Every scientist I spoke with stressed that overstating the impact threat is counterproductive, that scary headlines and bad information can eventually lead to less public support for asteroid research and planetary defense programs, not more. I hope that nothing in this book will be mistaken for hyping up the threat.

Still, the threat is real enough that we have some serious decisions to make. As the second decade of the twenty-first century comes to an end, we find ourselves in a window of opportunity. We can continue to build on the progress we've made so far, devoting the relatively modest resources and energy needed to fully understand Earth-threatening asteroids and develop planetary defenses against them. Or we can decide that because the risk of a catastrophic impact in our own lifetimes is so low, we will devote our attention to more pressing problems and, if the need arises, we will let someone else worry about asteroids—say, for example, our grandchildren.

Given what we've learned over the past two centuries, and especially during the past forty years, it should be an easy decision to make. And if you're still not convinced, take exit 233 off I-40, pony up the eighteen bucks for admission and take a good long look at Meteor Crater. It just might change your mind.

And now it's getting dark, and it's a long way home. But before I get on the freeway I pull my pickup to the side of the two-lane road and get out to spend a few minutes gazing up at the blackening sky. The Moon and the stars are there as they've always been, and I know if I watch long enough I'll see a flashing meteor or two.

But what I'm really looking at are the things that I can't see. Icy comets in the frigid far-off Oort Cloud and Kuiper Belt. Asteroids of iron and stone tumbling and turning in the main belt. Microscopic motes of cosmic dust gently wafting down through the atmosphere. Near-Earth Objects zipping past our planet at almost unimaginable speeds, some so close that it seems as if you can almost feel them. And who knows? Maybe one of them is destined to someday collide with our planet.

I can't see those things, but now I know they're out there.

And I know I'll never look at the sky in the same way again.

ACKNOWLEDGMENTS

It's been said that writing is a lonely business. But writing and publishing a nonfiction book is far from a solitary effort. There are many, many people without whom this book would not have been possible.

At Scribner, Editor in Chief Colin Harrison endorsed this project and Executive Editor Rick Horgan ramrodded it from the start; Rick kept me on track whenever I started to go astray. Editorial assistant Emily Greenwald handled the myriad details with enthusiasm and good humor; she's going to go far. Thanks also to production editor Jason Chappell and designer Kyle Kabel.

My agents, the redoubtable Eric Lasher and the inestimably talented Maureen Lasher of the L.A. Literary Agency, were the first to embrace this book idea and they never let go. Their friendship and wise counsel over many years has changed my life for the better.

I'm deeply grateful to the people who consented to be interviewed and quoted in this book. They were invariably patient and understanding in answering my questions—including the dumb ones, of which there were too many. Some of them graciously took the time to read portions of this narrative and to offer suggestions and make corrections—although of course any errors that remain are mine alone. They include Lindley Johnson, NASA;

Eric Christensen and Richard Kowalski of the Catalina Sky Survey; Bong Wie, Iowa State University; Gary Hug, Sandlot Observatory; and John S. Lewis, Deep Space Industries. Special thanks to David H. Levy, co-discoverer of Comet Shoemaker-Levy 9 and author of *Shoemaker by Levy*; and Donald K. Yeomans, former manager of JPL's Near-Earth Object Program Office and author of *Near-Earth Objects: Finding Them Before They Find Us*. Their books helped make this book possible.

Special thanks also to Lindsey Greytak of the University of Montana Russian language program for her meticulous translation of the Chelyabinsk videos, and to Jonathan Schleifer of Princeton University for his dogged research in the Barringer Family Papers at the Princeton University Library, Department of Rare Books and Special Collections. They were a pleasure to work with.

A number of people not named in the narrative were nonetheless instrumental in its production, either through interviews or providing other valuable assistance. For that I am grateful to Erin Morton, University of Arizona Lunar and Planetary Laboratory; Laurie Cantillo, NASA; Danielle Gunn, the Planetary Society; Lewin B. Barringer III, Barringer Crater Co.; Kenneth J. Zoll, Verde Valley Archaeology Center; Greg McKelvey, Gempressphotos.com; Peter Stibrany, Deep Space Industries; Dianna Ramirez, the Aerospace Corporation; Alexandria Bruner, FEMA; Nolan O'Brien, Lawrence Livermore National Laboratory; David Agle, JPL; Rob Seaman, Catalina Sky Survey; Betty Murphy, Heard Museum Library; Lanah Butterfield, Meteor Crater Enterprises; Hugh Miller, Paul Jones, and Agata Bogucka, Colorado School of Mines; Larry Lebofsky, Planetary Science Institute; Jeff Beal, Meteor Crater; Tim Swindle, University of Arizona Planetary Sciences Department; Dr. James B. Klein; Melissa Mead, University of Rochester, Rush Rhees Library; Theodore C. Tenny; and George F. Shaw.

Thanks also to my friend Ric Randall of Helena, Montana, who read portions of the manuscript and schooled me on the finer points of dinosaurs and uniformitarianism. Phil Garlington, Larry Wisocki, C. P. Smith, and Ruben Castaneda provided cheerful advice along the way. Thanks to Michael Wapstra for transportation services, and to Alex Wapstra and Betty Warren just for being there. I'm especially grateful to Isaiah Michael Weil and Emma Rose Weil, who helped me see the wonders of the universe through a child's eyes.

And finally, Debbie Weil and Annie Dillon gave me love, support and encouragement—as they always have.

My heartfelt thanks to all of you.

CHAPTER NOTES, SOURCES, AND RELEVANT FUN FACTS

INTRODUCTION

1 **just before 4 a.m.**: The precise time was 3:57 a.m. Mountain Standard Time. Most of Arizona does not observe Daylight Saving Time, even in June.

1 **the deadly volcanic explosion of Mount St. Helens**: On May 18, 1980, Mount St. Helens in Washington exploded in the most violent volcanic eruption in U.S. history, killing more than fifty people and sending more than 500 million tons of ash into the atmosphere. I was a young freelance writer living in Missoula, Montana, 500 miles to the northeast, and a few hours after the eruption I saw an ominously dark black cloud bearing down on the town, a cloud so dark and dense that it blocked the Sun and caused the automatic streetlights to come on in midday. Eventually the ground was covered in a quarter-inch of fine, talcum-powder-like volcanic ash that shut down the town for a week.

2 **exploded in the sky some fifteen miles above Arizona's White Mountains**: A group from Arizona State University later spent five days searching for remnants of the space rock in the rugged White Mountains on the Fort Apache Indian Reservation. Almost miraculously they managed to find fifteen small meteorites, from the

size of a pea to one the size of a strawberry. The stones are known collectively as the Dishchii'bikoh Ts'ilsoosé Tsee meteorite—that's Apache for Red Canyons Star Stone—and while the tribe retains ownership, by agreement they are being stored and studied at ASU's Center for Meteorite Studies.

CHAPTER 1 IMPACT!

9 **now extinct Late Pleistocene animals collectively known as mega-fauna:** When President Thomas Jefferson sent Meriwether Lewis and William Clark to explore the newly acquired Western territories in 1804 he urged them to be on the lookout for mastodons. Obviously they didn't spot any. There's continuing debate over whether ancient megafauna in North America went extinct because of climate changes or over-hunting by newly arrived humans.

11 **the energy equivalent of some twelve *million* tons of TNT:** Estimates of the energy released by the Meteor Crater asteroid impact are all over the map, from a mere 2.5 megatons all the way up to 60 megatons. I've used the estimate that seems to make the most sense. There are a number of interactive websites that allow you to calculate the destructive power of an impacting asteroid based on size, speed, composition, etc. Among them are www.purdue.edu/impactearth/ and https://impact.ese.ic.ac.uk/ImpactEarth/ImpactEffects/.

14 **by building the Standin' on the Corner municipal park:** The line from the Eagles song is "standin' on *a* corner" but Winslow decided to go with "*the* corner" instead.

16 **a free "Don't Forget Winona" postcard:** If you're of a certain age you might be interested to know that "(Get Your Kicks on) Route 66" was written by actor-songwriter Bobby Troup, best known as Dr. Joe Early on the 1970s TV show *Emergency!*

CHAPTER 2 MIRACULOUS APPARITIONS IN THE AYRE

23 **Dekker called them "thunder-stones":** Dekker didn't coin the term. In his 1608 play *Cymbeline*, William Shakespeare had this to say about rocks falling from the sky: "Fear no more the lightning flash, / Nor th' all-dreaded thunder-stone."

24 **Some of them reside in what's known as the Kuiper Belt:** The
Kuiper Belt is named after Dutch-American astronomer Gerard
Peter Kuiper. The Oort Cloud is named after Dutch astronomer
Jan Hendrik Oort.

25 **Like I said, it's far:** Beyond the Solar System, even astronomical units
are too small to handle the distances involved. To express distances to
other stars and galaxies we have to use light-years, one light-year being
the distance that light travels in 365 Earth days—that is, about six trillion
miles. But astronomers actually prefer "parsecs," which are equal to a
little more than three light-years each, or 19 trillion miles. There are
also "kiloparsecs," which are a thousand parsecs each; "megaparsecs,"
a million parsecs each; and "gigaparsecs," a billion parsecs each. Long
before you get to kiloparsecs, the layman mind begins to boggle.

27 **there should be a lot more comets:** Shakespeare weighed in on
comets, too. In his 1599 tragedy *Julius Caesar*, Caesar's wife, Calpurnia, tries to dissuade Julius from walking into almost certain
assassination at the Senate house in Rome. Describing various bad
omens, Calpurnia declares, "When beggars die, there are no comets
seen. The heavens themselves blaze forth the death of princes!" As
Shakespeare well knew, a spectacular Great Comet did appear over
Rome a few months after Caesar's assassination in 44 BC.

Of course, what comets foretell can depend on who's doing the
foretelling. One of the best examples is the Great Comet of 1066.
It's said that when the comet appeared over England and France,
with a glowing three-forked tail that covered half the sky, King Har
old of the Anglo-Saxons took the customary view that it presaged
disaster for him and his people; across his kingdom there were great
lamentations and gnashing of teeth. Meanwhile, on the other side
of the Channel, the Duke of Normandy saw the same comet and
reportedly proclaimed it "A wonderful sign from Heaven," prompting his soldiers to take up the cry, "A new star, a new king!" A few
months later, at the Battle of Hastings, both interpretations proved
true. King Harold was killed, reportedly by an arrow through the eye,
and the Duke of Normandy became King William the Conqueror.

28 **Halley predicted that the Great Comet of 1682 would reappear
after being gone for 76 years:** Those re-appearances actually occur
every 74 to 79 years, not the hard and fast 76 years Halley suggested.

29 **They are literally gone in a flash**: It's the brilliance and transitory nature of meteors that give them their romantic appeal. When someone advances rapidly in a given field, we often say that his or her rise was "meteoric"—although the word also connotes something that is brilliant but ends quickly. The writer Jack London, author of *The Call of the Wild* and other popular books, reportedly put it this way: "I would rather be a superb meteor, every atom of me in magnificent glow, than a sleepy and permanent planet." London got his wish. He died of kidney failure aggravated by alcoholism and morphine addiction at the relatively young age of forty. And no less an authority on meteoric rises and falls than Napoleon Bonaparte had this to say, no doubt referring to himself: "Men of genius are meteors destined to burn themselves out in lighting up their age."

31 **at which point they're called "meteorites"**: Like comets and meteors, meteorites occupied a special place in ancient cultures, both spiritually and on a practical level. Before the Iron Age, nickel-iron meteorites were the only significant source of that marvelous metal. The Inuit people in Greenland made harpoon points and other tools from a massive meteorite, and ancient Egyptians used meteoritic nickel-iron to craft elaborate ceremonial knives and daggers; one of them was found not long ago with the almost four-thousand-year-old mummified remains of the Egyptian pharaoh Tutankhamun, also known as King Tut. On the spiritual level, meteorites were venerated by people around the globe, for obvious reasons: Any large rock that fell out of the clear blue sky just had to have been sent by the gods. Some Meso-American cultures believed that meteorites were literally excreta from the gods—sweat, feces, semen—and small meteorites have been found in a 2,000-year-old Native American ceremonial altar in Ohio. The ancient Greeks and Romans both had religious cults centered around meteorites. One of the most infamous was founded by the mad teenaged Roman emperor Elagabalus—described by one historian as "the sorriest scapegrace ever to sit on a throne"— who had a black meteorite from Syria paraded through the streets of Rome in a golden chariot; spectators were strongly urged to cheer.

31 **hit something that went *clank*!**: The Hoba meteorite was discovered in 1920 when a white farmer named Jacobus Hermanus Brits was plowing a field in South West Africa—now Namibia—and the plow

blade got hung up on a hard object just below the surface. Brits dug down and found a nine-foot-wide, two-foot-thick, square-shaped nickel-iron meteorite; it looks sort of like a giant Wendy's hamburger patty. It apparently fell to Earth about eighty thousand years ago, so its impact certainly could have been witnessed by humans. The American Museum of Natural History in New York tried to buy it from the farm's owners in 1954, but calculations showed that the tracks on the nearest railroad line couldn't have borne its weight. It was declared a national monument in 1955, but over the years poachers chipped off hundreds or even thousands of pounds of the meteorite for sale on the private meteorite market. The Hoba meteorite is now better protected, and is one of Namibia's most popular tourist attractions.

32 **for which she received $80 and a carton of Cavalier cigarettes**: http://ivegotasecretonline.com/about/episode-guide/year-1954/.

34 **the missing planet in the Mars-Jupiter gap**: Even before Giuseppe Piazzi was invited to join the Celestial Police he was already on the case, spending his nights searching for the missing planet. The thousandth asteroid discovered, a 25-mile-wide main belt asteroid that was spotted in 1923, was named Piazzia in Giuseppe's honor.

35 **Those pieces are the asteroids**: In the early days of Solar System formation there were numerous "proto-planets" zipping around in the inner Solar System, some of them very large, and they often collided with one another. One theory holds that a proto-planet about the size of Mars smashed into Earth with such force that it sent huge pieces of the Earth's crust sailing off into space, where they eventually came together to form the Moon. That impact, or a series of other ones, may have caused the Earth to tilt about 23 degrees on its axis—which is why we have spring breaks and autumn harvest festivals. Other planets experienced world-altering collisions with proto-planets as well. For example, Uranus has an axial tilt of about 97 degrees, which means it rotates on its side, with its poles pointing toward the Sun. The prevailing wisdom is that it got hit by an Earth-sized proto-planet that pushed it into a sideways axis; you could say that as a result of a cosmic collision, Uranus got knocked on its ass. As for the name Uranus, when British astronomer William Herschel discovered the planet in 1781 he tried to name it Georgium

Sidus—Latin for "George's Star"—after King George III. But a German astronomer suggested the name Uranus, the Latinized form of Ouranos, the ancient Greek god of the sky. The Brits stubbornly stuck with Planet George for the next seventy years, but eventually Uranus carried the day—and it's been prompting sophomoric wisecracks ever since, like the one I just made above.

36 **most come in an endless variety of irregular shapes**: *Daily Show* host Trevor Noah had this to say about asteroid shapes. Asked in a 2018 CNN interview for his thoughts about having Donald Trump as president, Noah said, "It's almost like an asteroid is heading for the Earth—and it's shaped like a penis. I think I'm going to die, but I know I'm going to laugh." It's certainly possible for an Earth-threatening asteroid to be shaped like a penis.

38 **a quarter-mile-wide asteroid named 4581 Asclepius**: The discoverer was Dr. Henry Holt. *Los Angeles Times*, April 20, 1989.

38 **"National Near-Miss Day"**: https://nationaldaycalendar.com/national-near-miss-day-march-23/.

38 **And those were just the ones that were spotted**: https://watchers.news/2018/05/27/newly-discovered-asteroid-2018-ky2-flew-past-earth-at-0-78-ld/.

39 **"KISS YOUR ASTEROID GOODBYE!"**: *New York Post*, March 13, 1998.

40 **NASA reported getting up to 300 calls a week**: https://www.cnn.com/2012/12/21/living/apocalypse-by-the-numbers/index.html.

41 **a number/letter designation by the International Astronomical Union (IAU) that reveals the year and order of discovery**: The first number is the year of discovery, as in 1997 XF_{11}, which was discovered in 1997. The first letter corresponds to the half month in which it was discovered, the first half of January being "A," the second half of January being "B," the first half of February being "C," and continuing through the last half of December being "Y." The second letter in the sequence shows the order of asteroid discoveries within a given half-month period—"A" for the first one, "B" for the second one, and so on. So the first asteroid discovered in 1998 would be 1998 AA, the second 1998 AB, and so on. That worked okay when the number of asteroid discoveries was low, but as things picked up they had to add a number after the second letter.

42 **asteroid names**: For a full list of asteroid names see https://minor planetcenter.net/iau/lists/MPNames.html.

CHAPTER 3   ASTEROID MINERS

45 **Daniel Moreau Barringer**: The last name is pronounced with a hard "g," as in the ringer of a bell. (Author interview with Lewin B. Barringer III, great-grandson of Daniel M. Barringer and currently vice president of communications for the Barringer Crater Company, September 25, 2018.) In the style of the day Barringer was always referred to publicly as "D. M. Barringer" but most of his friends called him "Moreau." Quotes from Barringer throughout this chapter come from the Barringer Family Papers, Princeton University Library, Department of Rare Books and Special Collections; *A Grand Obsession: Daniel Moreau Barringer and His Crater* by Nancy Southgate and Felicity Barringer (Barringer Crater Co., 2002); and *Coon Mountain Controversies* by William Graves Hoyt (University of Arizona Press, 1987), p. 37. The late Mr. Hoyt's book is the definitive source on the history of Meteor Crater.

45 **he was largely self-taught**: An inventor of sorts, Barringer held two U.S. patents. One 1892 patent was for "a useful apparatus for illustrating Geological Formations," a set of interchangeable wooden blocks designed to show various types of ore beds and rock strata; it was sort of like a geological Rubik's Cube. His other patent was for a specially notched metallic front sight to be used on "rifles, guns and other firearms." Neither invention went anywhere. Barringer was also a writer, the author of two books and numerous articles. One of his books, *The Law of Mines and Mining in the United States*, written with a lawyer named John Stokes Adams and published in 1897, became the standard reference work on the subject—all 878 pages of it. His other book, *A Description of Minerals of Commercial Value*, was a short (82 pages) technical book that when you're reading it seems much, much longer. The complete subtitle gives a sense of the style: "A Practical Reference-Book for the Miner, Prospector, and Business Man, or Any Person Who May Be Interested in the Extraction or Treatment of the Various Metallic or Non-Metallic Minerals, and for Students Either in Field-Work or in the Laboratory."

46 **a small but potentially lucrative mining operation in southeastern Arizona**: The deposit was discovered in 1894 by a rancher and former hard-rock miner named Johnny Pearce. The story goes that Johnny was sitting on a hillside one day, enjoying a canned sardines sandwich, when he noticed an unusual rock on the ground. Cracking it open, his miner's eye saw that it was rich in traces of gold and silver. Pearce and his sons quickly staked a claim and started digging, sending the ore to a smelter in El Paso, Texas, but the mining operation was mostly small-time pick and shovel stuff. Barringer and his investors paid Pearce the then-enormous sum of $240,000 (about $7 million in today's dollars) for the mining rights, with $20,000 down and the rest taken out of future profits—if there were any. An odd clause in the contract required Barringer's company to build a boardinghouse at the mining site and allow Pearce's wife to run it. Mrs. Pearce apparently just enjoyed running boardinghouses; after the Commonwealth deal she certainly didn't need the money.

47 **individual named Samuel J. Holsinger**: Holsinger was in many ways one of the most interesting people in this story. I wish I could have given him his own chapter, but there are these people called editors whose job it is to move things along, so I'll just provide this thumbnail. Holsinger was trained as a lawyer, but he also worked as a rancher and newspaper reporter in California, including a stint as the court reporter for the *Los Angeles Herald*. (Ninety years later, as a young journalist, I had the same job at the same newspaper.) He was also a consumptive, or, pejoratively, a "lunger"—that is, he had tuberculosis. It was TB that led him to take the job as a "special land agent" for the General Land Office in 1897 and move to the drier and seemingly more healthful climes of Arizona. By all accounts Holsinger performed his duties honestly and seriously—too honestly and seriously for some people who, then as now, objected to the preservation of natural resources. Holsinger brought millions of new acres of land under the protection of the "forest reserve" system. Holsinger was also instrumental in obtaining federal protection for the Petrified Forest and the magnificent twelfth-century Native American ruins at El Chaco in New Mexico, among other natural and cultural sites. To sum it up, S. J. Holsinger was a key player in the preservation of some of the most beautiful and interesting

places in the American West—and everyone who has a National Parks pass should be grateful to him. As for his involvement in Barringer's potentially crater-destroying mining scheme, well, he had a wife and four children to provide for and not much time left—he died in 1911 at age fifty-two. For a thorough treatment of Holsinger's accomplishments, see "Preserving Our Western Natural and Historic Heritage: The Enduring Legacy of S.J. Holsinger" by Dr. James B. Klein, published in the *The Smoke Signal*, Tucson Corral of the Westerners, October 2008.

48 **it's also almost certainly bogus:** Charles Albert "Buckskin Charlie" Franklin, real name Albert Franklin Banta, was a scout for the Wheeler Survey, an ambitious multi-year effort by the Army Corps of Engineers to explore and map the American Southwest. Led by Lt. George Wheeler, the survey expedition consisted of a few dozen Army cavalrymen accompanied by civilian topographers, geologists, photographic specialists, and horse-wranglers and guides, of which Buckskin Charlie was one. According to Charlie, one day in 1873, "I saw a small hill about a mile to the eastward and went to it. Ascending to the top, to my surprise it was not a hill but a tremendous hole in the ground. . . . The more I looked across it the further away the other side seemed to be. I had no time to make an examination, but . . . I decided the distance across, from rim to rim, to be about one mile and a quarter, perhaps a little more, and so reported to Mr. Summers [a topographer with the survey group]. Mr. Summers made some memoranda and said, 'As you made the discovery, I shall name the hole for you and call it Franklin's Hole,' and so it appears in the records of the Wheeler expedition of 1873."

The problem with Charlie's story is that while it's true he worked as a guide for the Wheeler Survey, Lieutenant Wheeler's voluminous reports on the expedition make no mention of Franklin or his eponymous hole. And Charlie himself didn't mention it in a book-length manuscript he later wrote about his many Wild West adventures. In fact, it wasn't until almost half a century later, in 1918, long after Meteor Crater had become a well-known natural attraction, that Charlie first remembered his discovery of "Franklin's Hole." At the time, Charlie was pushing eighty, and he was living in the Arizona Pioneers Home in Prescott. (Other residents of the old-timers' home included "Big Nose Kate" Elder—Doc Holliday's old girlfriend—and a guy who claimed to have been Billy the Kid.) The Pioneers Home was the source of many inventive tales about the old days, and it certainly appears that Charlie's story about "Franklin's Hole" was one of them. Charlie's fanciful story is debunked in William Graves Hoyt's "Meteor Crater: Historical Note on Nomenclature" in the journal *Meteoritics* 18, no. 2 (June 30, 1983): 159–63.

49 **wrapped in a turkey feather blanket and interred in a ceremonial stone crypt:** Author interview with Kenneth J. Zoll, director of the Verde Valley Archaeology Center in Camp Verde, Arizona, July 14, 2017.

49 **they were worth good money:** Then as now, meteorites possessed both scientific and monetary value. And as with any valuable product, it wasn't long before Westerners started stealing them from native peoples who didn't grasp their true worth. There are many examples, but perhaps the most outrageous came in the late nineteenth century, when Arctic explorer and U.S. Navy officer Robert Peary hornswoggled a small band of Inuit people in Greenland. As mentioned above, for centuries the Inuit had used pieces of iron meteorites to make spear points and tools. Peary, who's been described by one historian as "probably the most unpleasant man in the annals of polar exploration," bribed an Inuit man to lead him to three large iron meteorites, the biggest of which weighed 36 tons. Over the course of several years Peary loaded the meteorites onto ships and took them to New York City, along with six Inuit people—then known as "Eskimos" or "Esquimaux"—for purposes of "scientific study." It all worked out quite well for Peary, who eventually sold the giant meteorites for $40,000 (about $1 million today) to the American Museum of Natural History, where they remain today. Things went considerably less well for the Inuit, however. Housed in the basement of the museum, four of them died within a year from heat and European diseases, while one returned to Greenland. The only other survivor was a seven-year-old boy named Minik, who grew up in New York City, far from his people and culture; he once described himself as "the loneliest person in the world." Minik became increasingly bitter about the treatment of his people by Peary and other Americans, calling them "a race of scientific criminals"—

for which you can hardly blame him. Peary later went on to win fame as the first man to reach the North Pole in 1909—although that claim was disputed—while poor Minik became a lumberjack in New Hampshire and died in the 1918 Spanish flu epidemic.

That's just one example. You can argue that preserving a meteorite in a museum or laboratory or even a private collection is better than having it slowly erode away on the ground. That's the point that modern-day meteorite hunters and meteoritic scientists make, and in some ways it's persuasive. Still, it's undeniable that some of the most famous meteorites in the world's most prestigious museums are stolen goods.

49 **the sheep-camp cook:** The cook was a down-on-his-luck prospector named Frederick Krapf—sometimes rendered as "Craft"—and he was a somewhat unusual figure. The story is that he wandered into the sheep camp one day, bruised and bloodied from being beaten up when he tried to strangle the bartender at the Bucket of Blood saloon in nearby Holbrook. He claimed to have been educated at the finest schools in Heidelberg, and to have graduated from the famous university there, although exactly how he wound up manning the chuck wagon in a dusty Arizona sheep camp is unclear. Krapf also had unusually long arms and hairy hands—so much so that they figured prominently in a contemporary account of him, which used the phrase "gorilla-like" to describe them. Maybe that's why it had been so hard to pry him off the bartender.

50 **including one 200-pounder:** Foote decided the meteorites were probably pieces of a five- or six-hundred-pound mass of space iron—an object he described as "extraordinarily large"—that had exploded high above the ground. The fact that Dr. Foote thought a 600-pound meteoroid was extraordinarily large illustrates just how little was known about large Earth-impacting space bodies in 1891. We now know that meteoroids that big or bigger routinely burn up or explode in the Earth's atmosphere—so a mere 600-pounder is neither extraordinary nor large.

50 **Chief among them was one Fred W. Volz:** A slight, dapper-looking man, Volz was licensed by the federal government to trade with the local Hopi and Navajo tribes, exchanging provisions and other necessities for hides, pelts, Navajo blankets, and so on, and by most

accounts he treated them fairly, at least by the standards of the day. Although he was known to occasionally fly into terrifying Teutonic rages when sufficiently lubricated by whiskey—or perhaps schnapps—ordinarily he was a genial sort. As for the town of Canyon Diablo, it started off as a canvas and tar paper shantytown inhabited by perhaps a thousand people, many of them railroad workers but including a significant population of prostitutes, pimps, gamblers and other hard cases. All sorts of lurid tales have grown up around Canyon Diablo. It's said that the town's Boot Hill was quickly stocked with thirty-five graves, all of whose occupants died by gunshot or blade or bludgeon, and that dozens of others lay about in shallow unmarked graves. As for law enforcement, the oft-told story is that five town marshals were hired and subsequently gunned down in rapid succession, one of them having been sworn in at 3 p.m. and buried before sundown the same day. And so on. Like so many Wild West tales, those stories have certainly grown in the telling.

51 **Canyon Diablos wound up stored in barrels**: Some pieces of the Canyon Diablo meteorite also wound up at the Paris laboratory of a French chemist named Henri Moissan, who in 1893 discovered that the meteorite contained tiny particles of silicon carbide, a mineral that was later named "moissanite." Moissanite is extremely rare on Earth, but there's a lot of it in space; it is, literally, "stardust" that is older than the Solar System. Almost as hard as diamonds, and with even greater sparkle qualities, industrially synthesized moissanite is now used as a much cheaper "diamond alternative" in jewelry. In fact, unless you're an expert it's hard to tell a one-carat, $5,000 diamond from a similarly sized $700 piece of moissanite—which makes you wonder how much of Henri Moissan's re-created Canyon Diablo meteorite discovery is encircling the fingers of unsuspecting fiancées. Among many other applications, lab-produced silicon carbide is also used in the protective ceramic plates in military body armor.

52 **"perhaps the closest equivalent to a saint"**: Hoyt, *Coon Mountain Controversies*, p. 37.

53 **Gilbert eventually mounted an expedition to the Arizona crater**: Although Gilbert doesn't specifically mention it in his handwritten field notes, the expedition's gear almost certainly included rifles and ammunition—and not just to provide meat for the pot. It's true that

by 1891 Arizona had in many ways been modernized. There were the railroads, of course, and telegraph lines and electric lights and local and long-distance telephone service in the bigger towns; Tucson had an opera house that could hold close to a thousand people. But most of the territory remained almost as wild and dangerous a place as it had been when Gilbert first visited it two decades earlier as a member of the aforementioned Wheeler Survey—and this time he didn't have a U.S. Cavalry escort. Just a few months before Gilbert arrived, African-American "Buffalo Soldiers" of the U.S. 10th Cavalry fought what turned out to be the last battle of the so-called Apache Wars, a running skirmish with a band of hostile Apaches in the eastern part of the territory. Not far from Coon Mountain, sheepmen and cattlemen were still waging the so-called Pleasant Valley War, a ten-year feud that saw fifty men lynched, stabbed or shot, usually in the back. Cattle rustling was a major enterprise, along with stagecoach and freight line and train robberies. There were four-legged dangers as well, with wolves posing a threat to livestock, and grizzly bears in the mountains a threat to both stock and men. The point is that while searching for an ancient asteroid in Arizona in 1891, it was wise to be well armed.

53 **"I'm going to hunt for a star"**: Today it seems a little strange for a scientist to have used the word "star" for an asteroid or meteoroid, but like "shooting stars" and "falling stars" it was a common usage back then. The words "meteor," "comet," and "asteroid" were often used interchangeably in those days, not only by journalists but by many scientists as well.

55 **that asteroid would be one of the most lucrative mining operations in the history of the world**: Barringer wasn't the first to envision fantastic wealth contained in asteroids. In an 1898 science-fiction novel titled *Edison's Conquest of Mars*, astronomer Garrett P. Serviss envisioned President Grover Cleveland and other world leaders dispatching an expedition led by Thomas Edison to wreak vengeance on Mars for a Martian attack on Earth. Along the way they encounter an asteroid made entirely of gold nuggets.

57 **"The past is a foreign country"**: The quote is from British novelist L. P. Hartley's book, *The Go-Between* (1953).

57 **there was some discussion about a bribe being involved**: Barringer directed one of his partners, E. J. Bennitt of Phoenix—he

was Barringer's brother-in-law—to file an expedited mining patent application with the federal Surveyor General's office in Phoenix. As a veteran federal agent, Holsinger could have warned him what would happen. "He [Bennitt] has been among the Philistines," Holsinger later reported to Barringer. "It was rather unfortunate that he conveyed to the Surveyor General the idea that [we] are desirous of securing an early patenting of our claims. The fact is that they will hold us up just in proportion to the amount of anxiety displayed by us. Plainly speaking, they are a set of robbers and bleed every man they can. If they think we want a patent badly they will hold us up for at least $1,000. The chief clerk, Mr. Murphy, has made a small fortune in that business. I came near catching these fellows several times and made them drop several hold-ups. . . . Of course, you may deem it expedient to pay something in order to ease this application through, but as I look at it we are engaged in a perfectly legitimate business and we ask for nothing we are not entitled to. I never took a bribe in my life, and I will have to know that there is no other way out of it before I will consent to give one." Whether any palms were actually greased is unknown.

60 **"a meteor which must have been of great size"**: D. M. Barringer, "Coon Mountain and Its Crater," *Proceedings of the Academy of Natural Sciences of Philadelphia*, March 6, 1906. One of Barringer's partners, an amateur mathematician and geologist named Benjamin Chew Tilghman, also submitted a scientific paper to the academy. Tilghman should not to be confused with the Benjamin Chew Tilghman who was a noted Civil War general and the inventor of the industrial sandblaster; this Tilghman was a nephew.

60 **forty-eight thousand years off the mark**: The 700-year minimum age for the crater came from Holsinger, who had counted the annular rings in a dwarf juniper tree found on the crater rim. Since no tree could have survived the asteroid impact, the crater had to be older than the tree. Barringer's 2,000-year maximum age was basically just a guess based on the amount of surface erosion.

62 **"My dear Barringer"**: Southgate, *A Grand Obsession*, p. 34.

66 **two groundbreaking books**: *Mining the Sky: Untold Riches from the Asteroids, Comets, and Planets* by John S. Lewis (Perseus Books, 1996), and *Asteroid Mining 101: Wealth for the New Space Economy* (Deep

Space Industries, 2014). Lewis is also the author of *Rain of Iron and Ice: The Very Real Threat of Comet and Asteroid Bombardment* (Helix Books, 1996).

67 **"Beam me up, Scotty"**: Actually, that exact phrase was never spoken by Capt. James Kirk in any episode of *Star Trek*.

67 **"People who want a piece of the action are flocking in"**: Author interview with John S. Lewis, May 14, 2018. Professor Lewis has a vested personal interest in the economy of the future. He has thirty-four grandchildren.

68 **"It will open a new era of space exploration"**: Chris Lewicki, Planetary Resources CEO, www.youtube.com/user/PlanetaryResources 2017.

68 **Every year a mere 200 tons of platinum worth about $6 billion are mined on Earth**: https://crosscut.com/2018/03/redmond-companys -cosmic-gamble.

71 **"Earth is being threatened by huge piles of money!"**: https://www .theguardian.com/science/2018/jun/09/asteroid-mining-space -prospectors-precious-resources-fuelling-future-among-stars.

72 **little confidence that private companies would try to minimize space pollution**: http://www.pewinternet.org/2018/06/06/majority -of-americans-believe-it-is-essential-that-the-u-s-remain-a-global -leader-in-space/.

73 **dedicated Meteor Crater solely to tourism and scientific research**: In 1982 the family-owned Barringer Crater Company, owners of the crater, established the Barringer Medal to recognize outstanding work in the field of impact crater research. The first recipient was Eugene Shoemaker. The Barringer Family Fund for Meteorite Impact Research also provides grants to graduate students to conduct field research at impact craters around the world. Meteor Crater has also provided a backdrop for a number of movies, including the 1984 Jeff Bridges film *Starman*. Meteorite fans will also note—perhaps with annoyance—that I've left Harvey H. Nininger out of the narrative. Nininger was a former college professor who in the 1920s began a lifelong search for meteorites and the study of their origins. The author of numerous books and articles, he founded the American Meteorite Museum, which for a time was located near Meteor Crater, and was considered perhaps the world's foremost authority on

the subject. I've left him out because although he sometimes wrote about large cosmic impacts, his primary interest was meteorites, not asteroids and comets. And as I said, there are these people called editors.

CHAPTER 4 STAR WOUNDS

76 "All villagers were stricken with panic and took to the streets": Russian newspaper *Sibir*, quoted in *Comets!: Visitors from Deep Space* by David J. Eicher (Cambridge University Press, 2013), p. 57.

78 "one of the most appalling catastrophes in human history": Unnamed Russian newspaper, quoted in the *New York Times*, December 2, 1928.

79 possibly changed the course of twentieth-century history: The "what ifs" surrounding the Tunguska Event have become something of a parlor game over the years. A number of people have suggested that if the asteroid/comet had collided with the atmosphere on the same trajectory but just a few hours later in the Earth's rotation cycle, it would have hit St. Petersburg, possibly killing a revolutionary named Vladimir Ilyich Ulyanov, also known as Lenin, with history-changing results. The only problem with that theory is that Lenin had fled St. Petersburg in 1907 and moved to Finland, and later Geneva (https://www.marxists.org/archive/lenin/bio/timeline .htm). Also, if the asteroid/comet's path had been just a little slower, Earth's 67,000-mile-per-hour orbit around the Sun would have taken it out of the asteroid/comet's path and it would have missed us altogether. It's fun to speculate, but "what ifs" don't really mean much. As the old saying goes, "If 'ifs' and 'buts' were beers and nuts, we could throw ourselves a party."

81 quickly shouted down by other physicists: For a detailed look at the various Tunguska theories, see Roy A. Gallant's entertaining *The Day the Sky Split Apart: Investigating a Cosmic Mystery* (Atheneum, 1995).

84 pressure of an asteroid impact: Later researchers uncovered a 5,000-pound meteorite at the Wabar site.

84 impact craters just outside of Odessa, Texas: Brandon Barringer, "Historical Notes on the Odessa Meteor Crater," *Meteoritics* 3, no. 4 (1967): 161–68.

84 **small metallic asteroid**: Daniel Barringer tried to buy the Odessa property and dig up the asteroid, but the railroad that owned it wouldn't sell—which was probably fortunate for Barringer's finances, since later drilling showed that the asteroid no longer existed, that like the Meteor Crater asteroid it had been mostly vaporized on impact. Today the Odessa Meteor Crater is a privately owned National Natural Landmark and tourist attraction, but at only about fifteen feet deep it's a visually poor cousin to Meteor Crater 700 miles farther west. The price is right for admission, though—it's free—but donations are welcomed.

88 **Fornicating man-bats!**: For an entertaining look at the Great Moon Hoax, see Matthew Goodman's *The Sun and the Moon: The Remarkable True Account of Hoaxers, Showmen, Dueling Journalists, and Lunar Man-Bats in Nineteenth-Century New York* (Basic Books, 2008).

89 **"questions about the Moon only in jest"**: Astronomer J. F. Julius Schmidt, quoted in Hoyt, *Coon Mountain Controversies*, p. 21.

89 **"gaping at the Moon"**: Gilbert didn't respond publicly to the congressman's attack; that wasn't his style. But later, after noting that his telescopic lunar observations were often hampered by cloud cover, he confided to a friend that "Clouds and congressmen are about equally obstructive."

90 **the impact theory of lunar craters**: Among other accomplishments, Robert Dietz also helped prove the theory of plate tectonics or "continental drift," and helped design the bathyscaph *Trieste* that in 1960 dived down seven miles into the Pacific's Challenger Deep. As for Ralph Baldwin, although he had a PhD in astrophysics, his day job when he wrote *The Face of the Moon* was as an executive with the Oliver Machinery Co. of Grand Rapids, Michigan, a manufacturer of woodworking machinery.

90 **Gene Shoemaker**: The best biography of Gene Shoemaker is by his friend and colleague, astronomer David H. Levy, *Shoemaker by Levy: The Man Who Made an Impact* (Princeton University Press, 2000).

91 **friends started calling him "SuperGene"**: This was a bit of nerdy Caltech humor. In geology, mineral-producing "hypogene" processes occur deep within the Earth, while "supergene" processes occur near the Earth's surface.

91 **the expected surge in civilian nuclear power plants**: Faced with the enormous costs of building nuclear weapons and an incipient Ban the Bomb movement, in the 1950s the civilian Atomic Energy Commission was pushing the peaceful uses of nuclear power. The AEC glowingly predicted that nuclear-generated electric power would soon be so cheap that power companies wouldn't even have to meter it. Today there are about one hundred commercial nuclear power reactors in the U.S.—and have you checked your electric bill lately?

92 **Meteor Crater was "pretty much just a scaled-up version" of the nuclear bomb craters**: One of the nuclear bomb craters Shoemaker studied was made by "Jangle U," a 1951 test to measure the effects of a ground-penetrating 1.2-kiloton bomb; it left a crater 260 feet wide and 53 feet deep. The other crater was from "Teapot Ess," in 1955, which involved an underground detonation of another 1.2-kiloton atomic bomb. It left a crater 300 feet wide and more than 100 feet deep. Both test series included positioning U.S. military troops near the explosion point to assess post-detonation combat effectiveness and the physical and psychological effects. This was just one example of the cavalier attitude toward radiation at the time. I can remember when shoe stores used X-ray fluoroscopes to see how shoes fit your feet.

93 **Shoemaker and two colleagues**: They were Beth Madsen and Edward C. T. Chao. Coesite was named after Loring Coes Jr., a private -industry researcher who was working on the synthetic production of diamonds under high heat and pressure.

94 **NASA shot Alan Shepard into a sub-orbital flight and put John Glenn into orbit—albeit well behind the Russians**: The Russians managed to send the first man into space—Yuri Gagarin orbited the Earth in 1961—but he wasn't the first life-form to be rocketed beyond the atmosphere. In 1947 the U.S. sent a capsule containing fruit flies 68 miles above the Earth, just beyond the space line, in order to test the effects of radiation exposure at high altitudes. The fruit flies in their capsule successfully parachuted back to Earth and survived the trip.

95 **"my biggest disappointment in life"**: Malcolm Browne, *New York Times*, July 22, 1997.

95 **"one small step for a man" in 1969**: Armstrong was quoted as saying "That's one small step for man, one giant leap for mankind," without

the "a" before "man"—which really doesn't make much linguistic sense. Armstrong later said the "a" was lost in transmission, which it may have been, or maybe in the excitement of the moment he understandably flubbed the line. Unfortunately the "a" is still missing in many descriptions of the first Moon landing.

95 **providing live commentary during some of the Apollo Moon missions:** Strangely enough, the Apollo program didn't necessarily make Shoemaker happy. From the beginning there had been an ongoing battle between NASA engineers and scientists like Shoemaker. The engineers just wanted to put humans on the Moon within the Kennedy deadline and before the Russians did it; the scientists wanted the missions to focus more time and effort on scientific exploration, not just getting there first. To Shoemaker's frustration, the engineers generally won out. In a 1969 magazine article Shoemaker publicly slammed NASA for "focusing on getting [a man] to the Moon . . . and not what he was going to do after he got there." Later, after NASA canceled the last couple of Apollo Moon missions, Shoemaker added, "The main issue for me was not flags or footprints [on the Moon]. . . . Had Apollo succeeded, we'd still be there exploring." Shoemaker's caustic comments caused some hurt feelings at NASA, and he generally separated himself from the space agency—although, Shoemaker being Shoemaker, he eventually would be warmly welcomed back.

95 **did more than just beat the Russians:** In addition to putting the first man in space the USSR also managed to decisively beat the USA in another endeavor—that is, putting women in space. In 1963 the Russians launched 26-year-old Valentina Tereshkova into orbit aboard a space capsule that circled the Earth 48 times over the course of three days before it re-entered the atmosphere and Tereshkova parachuted out. (That was how Russian cosmonauts landed in those days.) It wasn't until twenty years later, in 1983, that American astronaut Sally Ride became the first American woman in space aboard the space shuttle *Challenger*.

96 **The Moon is just 240,000 miles away from Earth:** Actually the Moon is getting a little farther away from the Earth with each passing day. For various astrophysical reasons, it's receding away from Earth at the rate of about an inch and a half a year.

96 **Mars alone has more than 600,000 impact craters bigger than a half mile across:** https://www.space.com/16153-mars-impact-crater-map.html.

97 **"It's like being in a hail of bullets going by all the time":** Shoemaker interview, https://www.springfieldspringfield.co.uk/movie_script.php?movie=national-geographic-asteroids-deadly-impact.

100 **professionally separated from Helin:** The split with Helin was bitter, at least on her end. She accused Shoemaker of taking credit for other people's work and of trying to undercut her role in the Palomar asteroid survey—charges that Shoemaker and his friends strongly denied.

102 **could devastate a city:** Curtis Peebles, *Asteroids: A History* (Smithsonian Institution Press, 2000), pp. 209–13.

103 **"to get the world to take notice of the threat of asteroid impact":** Speech by Brig. Gen. Simon Worden, "Military Perspectives on the Near-Earth Object (NEO) Threat," released by United States Space Command, Monday, July 15, 2002.

CHAPTER 5 T-REX WITH A STRING OF PEARLS

106 **deposited roughly sixty-five million years ago:** As usual, there is a debate over whether the extinction occurred 65 million years ago or, as some researchers say, 66 million years ago. The 65-million-year figure is the one most often used in popular culture, so I'll stick with that.

106 **one of the greatest mass extinctions in Earth's history:** Scientists being scientists, they have to make things complicated. What was formerly called the K-T or Cretaceous-Tertiary Boundary—the K is from the German word for "chalk"—is now called the "Cretaceous-Paleogene Boundary," or "K-Pg Boundary."

107 **"paleoweltschmerz," a disenchantment with life in a difficult world:** Merriam-Webster defines "weltschmerz" as "mental depression or apathy caused by comparison of the actual state of the world with an ideal state." Literally it translates from the German as "world pain."

108 **"Nature seemed to be showing us something quite different":** Walter Alvarez, *T. rex and the Crater of Doom* (Princeton University Press, 1997), p. 42.

108 **its mission to Hiroshima:** Believe it or not, the pilots' call sign for the Hiroshima mission that killed a hundred thousand people was "Dimples."

108 **the CIA-sponsored Robertson Panel that investigated alien UFOs in the early 1950s:** The Robertson Panel, named after its chief, physicist Howard P. Robertson, concluded that "the evidence presented on Unidentified Flying Objects shows no indication that these phenomena constitute a direct physical threat to national security. We firmly believe that there is no residuum of cases which indicates phenomena which are attributable to foreign artifacts capable of hostile acts, and that there is no evidence that the phenomena indicates a need for the revision of current scientific concepts." The panel added that "the continued emphasis on the reporting of these phenomena does, in these parlous times, result in a threat to the orderly functioning of the protective organs of the body politic. We cite as examples the clogging of channels of communication by irrelevant reports, the danger of being led by continued false alarms to ignore real indications of hostile action, and the cultivation of a morbid national psychology in which skillful hostile propaganda could induce hysterical behavior and harmful distrust of duly constituted authority." Predictably, after the panel's top-secret report was declassified it was widely viewed as a cover-up.

111 **no one paid attention:** The earlier suggestions of impact-related extinctions seem prophetic in retrospect. But we shouldn't make too much of them. It's a little like an investigation that takes place after a terrorist attack. You can always find obscure references in the mountains of intelligence "chatter" that could have been related to the attack, but were too vague to act upon at the time. Again, the Alvarezes were the first to present physical evidence of an extinction-level impact.

112 **the global effects of that massive impact had caused the K-T extinction:** The K-T extinction was just one of five major extinction events in Earth's history, along with a number of smaller ones. The most destructive to life was the Permian-Triassic extinction event of 250 million years ago, which scientists believe wiped out 96 percent of all species on Earth. Luis Alvarez believed that all five major extinction events had been caused by cosmic impacts. Maybe so, but the scientific evidence is slim.

113 **scale-model Tokyo in** *Godzilla*: The Godzilla suit weighed some 200 pounds and was so stifling that the stunt man could wear it for only a few minutes before passing out. As for the film *One Million Years B.C.*, the premise that humans and dinosaurs once shared the Earth is of course preposterous. But how can science compete with Raquel Welch in a fur bikini?

113 **a green** *Apatosaurus* **on the company logo**: Sinclair first used the "Dino the Dinosaur" logo on cans of motor oil in 1930, with the message that the oil had been "Mellowed for 100 Million Years." Dino was originally described as a *Brontosaurus*, but Sinclair later adopted new scientific theories and called it an *Apatosaurus*. Either way, it's still part of the company logo.

114 **"Astronomers should leave to astrologers the task of seeking the cause of Earthly events in the stars"**: "Miscasting the Dinosaur's Horoscope," *New York Times*, April 4, 1985.

115 **"some kind of scam"**: Charles Officer and Jack Page, *The Great Dinosaur Extinction Controversy* (Basic Books, 1996), pp. xiii and 185.

116 **"Had she taken a bullfighter"**: Bill Bryson, *A Short History of Nearly Everything* (Broadway Books, 2003), p. 137.

116 **"They're more like stamp collectors"**: *New York Times*, January 19, 1988. The stamp collector put-down was a common theme among physicists. The early-twentieth-century nuclear physicist and Nobel Prize winner Ernest Rutherford supposedly once said, "All science is either physics or stamp collecting"—a slam that has been repeated often since. Bryson, *A Short History of Nearly Everything*, p. 137.

117 **"Paleoenvironments of the Vertebrate-Bearing Strata During the Cretaceous-Paleocene Transition"**: Today Google Scholar lists some 16,000 scientific books and papers related to the Alvarez impact-extinction theory.

118 **village of Chicxulub**: "Chicxulub" is Mayan for either "place of the cuckold" or "the red devil." James Lawrence Powell, *Night Comes to the Cretaceous* (W. H. Freeman, 1998), p. 103.

120 **Palomar Asteroid and Comet Survey**: The Palomar Asteroid and Comet Survey (PACS), 1983–1993, Authors: Shoemaker, C. S., Holt, H. E., Shoemaker, E. M., Bowell, E., & Levy, D. H., Abstracts for the IAU Symposium 160: Asteroids, Comets, Meteors, 1993. Held June 14–18, 1993, in Belgirate, Italy.

121 "it looks like a squashed comet": Levy, *Shoemaker by Levy*, p. 10.

121 "we'd found a real unicorn!": David H. Levy, *Impact Jupiter: The Crash of Comet Shoemaker-Levy 9* (Plenum Press, 1995), p. 27.

123 "I don't believe it": Levy, *Shoemaker by Levy*, p. 219.

124 "Cosmic Crash": *Time*, May 23, 1994.

124 "Comet to Hit Jupiter with Texas-Sized Bang": *New York Times*, October 19, 1993.

124 Jupiter was thoroughly covered: The predicted Shoemaker-Levy 9 impact also produced no end of silliness. For example, an English "mystic astronomer" named Sofia Richmond, aka Sister Marie, declared that Shoemaker-Levy 9 was actually Halley's Comet making an abrupt return after its 1986 apparition and mounting an attack on Jupiter that would blow up the entire planet. Sister Marie took out full-page ads in several British newspapers announcing "A Warning Ultimatum to All Governments— SOS to the Pope," and advising that Earth could save itself from collateral damage by abolishing alcohol and pornography and all manner of crime. She also urged people who watched the explosive collision to wear sunglasses.

125 "Comet Crashes into Jupiter in Dazzling Galactic Show": *New York Times*, July 16, 1994.

125 "Comet Scars Jupiter with Earth-sized Blot": *Washington Post*, July 19, 1994.

125 "Everyone knows that 'fizzle' is a Yiddish word": Levy, *Impact Jupiter*, p. 268.

126 a network eventually known as "Spaceguard": The Spaceguard name came from a science-fiction novel by Arthur C. Clarke, author of *2001: A Space Odyssey* and numerous other popular sci-fi stories. In the 1973 novel *Rendevouz with Rama*, Clarke described an asteroid wiping out several cities and six hundred thousand people in Italy in the year 2077, a human and cultural catastrophe that prompted the world to set up an asteroid-defense network. As Clarke put it, "No meteorite large enough to cause catastrophe would ever be allowed again to breach the defenses of Earth. So began Project SPACEGUARD." Arthur C. Clarke, *Rendezvous with Rama* (RosettaBooks, 2012, Kindle Edition).

128 "the epoch of large asteroid strikes on Earth ended millions or billions of years ago": Robert Park, Lori Garver, and Terry Dawson, *Hazards Due to Comets and Asteroids*, edited by Tom Gehrels (Uni-

versity of Arizona Press, 1995). Quoted in Carrie Nugent's *Asteroid Hunters* (TED Books, 2017), p. 80.

128 **"Ask any dinosaur, if you can find one. This is a dangerous place"**: Timothy Ferris, "Is This the End?" *The New Yorker*, January 27, 1997.

129 **"When Worlds Collide: A Threat to Earth Is a Joke No Longer"**: William J. Broad, *New York Times*, August 1, 1994.

129 **"Jupiter this week has changed a lot of minds"**: "The Doomsday Asteroid," *NOVA*, PBS, April 29, 1997, http://www.pbs.org/wgbh /nova/transcripts/2212doom.html.

130 **scientifically it was execrable**: *Armageddon* co-star Ben Affleck later said that after reading the initial script he asked director Michael Bay why it was easier to teach oil drillers how to be astronauts than it was to teach astronauts how to be drillers—at which point, Affleck said, "He [Bay] told me to shut the fuck up." https://filmschoolrejects .com/61-things-we-learned-from-the-armageddon-commentary -426b81c04fbc/.

132 **only time that a human being had ever been interred on another celestial body**: The comet pictured on the capsule was Hale-Bopp, and the lines from *Romeo and Juliet* were:

> And, when he shall die,
> Take him and cut him out in little stars,
> And he will make the face of heaven so fine
> That all the world will be in love with night,
> And pay no worship to the garish sun.

Today you can send a tiny portion of your loved one's cremated ashes—or by pre-arrangement, your own—into space via several private companies. For example, a Houston-based company called Celestis will arrange to send ashes into Earth orbit for about $5,000; ashes from *Star Trek* creator Gene Roddenberry and American psychologist and LSD expert Timothy Leary were sent into orbit two decades ago. Celestis plans to send ashes to lunar orbit or a Moon landing at a future date for $12,500. As of March 2018 they were still accepting reservations.

136 **radio telescope imagery plays a big role in that**: Another important asteroid-imaging radar system is the Goldstone Solar System Radar

facility, located in the desert near Barstow, California. Its dishes are smaller than Arecibo's, but they've provided some really good asteroid images over the years, including a stunningly detailed look at our old friend asteroid 4179 Toutatis as it passed within four million miles of Earth in 2012.

CHAPTER 6 ASTEROID HUNTERS

139 **"There's one," Kowalski says**: Author interview with Richard Kowalski, March 12, 2018.

140 **You might not think that astronomers as a group have a great sense of humor**: Kowalski was featured in a May 11, 2009, article in the satirical website *The Onion*, headlined "Chicken-Shit Asteroid Veers Away at Last Minute." The article reported, "Though initial calculations showed it to be on a direct collision course with Earth, a pansy-ass asteroid approximately the size of Rhode Island has instead altered its trajectory to avoid the planet by more than 40,000 miles, astronomers at the University of Arizona reported Monday. 'Guess it just didn't have the spuds to go through with it,' Richard A. Kowalski of the school's Catalina Sky Survey said. 'Real big surprise. Maybe you can try again when you accrete a little more mass than 6.32×10^{15} kilograms, okay? Chicken-shit.'" Of course, Kowalski had nothing to do with the *Onion* story. But he thought it was pretty funny.

146 **A European service called NEODyS**: The Near-Earth Objects Dynamic Site is a European Space Agency–supported program. It's operated by the University of Pisa in Italy and the University of Valladolid in Spain.

146 **The Torino Scale defines the impact hazard**: Designed by Richard P. Binzel of MIT, the scale was adopted during a 1999 NEO conference held in Turin, Italy. We call it Turin, the Italians call it Torino, just like we say Rome and Italians say Roma. In deference to local wishes, Torino Scale, not Turin Scale, carried the day.

149 **even from an amateur observatory on the campus of New Milford High School in Connecticut**: Nugent, *Asteroid Hunters*, p. 31.

149 **"That was an interesting day"**: Author interview with Paul Chodas, JPL, May 24, 2018.

153 "Our time is a lot cheaper than theirs is": Author interview with Gary Hug, April 17, 2018.

156 "It's not really a competition": Author interview with Eric Christensen, April 24, 2018. By the way, although he has always worked in astronomy, Christensen graduated from the University of Arizona with a bachelor of fine arts degree, with an emphasis on ceramic sculpture.

158 only a tiny portion of NASA's $18 billion budget that year: *Christian Science Monitor*, January 14, 2016, https://www.csmonitor.com /Science/2016/0114/NASA-creates-Planetary-Defense-Coordination -Office.-Why-now.

160 said it was important for NASA to continue its search for potentially dangerous asteroids: http://www.pewinternet.org/2018/06/06 /majority-of-americans-believe-it-is-essential-that-the-u-s-remain-a -global-leader-in-space/.

160 By comparison, in 1993: Clark R. Chapman, "History of the Asteroid/Comet Impact Hazard," Southwest Research Institute, revised, October 7, 1998, http://www.boulder.swri.edu/clark/ncarhist.html.

CHAPTER 7 PLANETARY DEFENSE

164 prompted a flurry of newspaper headlines: "Asteroid to Buzz Earth Next Week," National Public Radio, https://www.npr .org/2013/02/08/171412450/close-shave-asteroid-to-buzz-earth-next -week; "Earth to Narrowly Escape Collision with Asteroid," *Washington Post*, February 14, 2013; "Talk About Close! Asteroid to Give Earth Record-Setting Shave," NBCNews.com, February 1, 2013, http://www.nbcnews.com/id/50672151/ns/technology_and_science -space/t/talk-about-close-asteroid-give-earth-record-setting-shave/# .Wtn5eUxFzIU.

165 sirens and car alarms wailing: https://www.youtube.com/watch?v= dpmXyJrs7iU.

165 "It was a light which never happens in life": *New York Times*, February 15, 2013.

167 "Surprise Attack: Meteor Explodes Over Russia": *Washington Post*, February 15, 2013; "Shockwave of Fireball Meteor Rattles Siberia," *New York Times*, February 15, 2013.

167 **"it was Americans testing their new weapons"**: https://www
.theatlantic.com/international/archive/2013/02/russian-meteorite
-conspiracy-theories-debunked/318293/.

169 **"Houston, we have a problem"**: Apollo 13 commander Jim Lovell
actually said, "Houston, we've had a problem," but in the film *Apollo
13* Tom Hanks says, "Houston, we have a problem," and that's the
version that stuck.

170 **Asteroid Day is now observed**: Resolution adopted by the General
Assembly on December 6, 2016.

171 **"it's not what I would call particularly high stress"**: Author interview
with Lindley Johnson, February 14, 2018.

173 **So the Air Force launched a study of future roles and missions**: Spacecast 2020, U.S. Air Force, 1993, http://csat.au.af.mil/2020/intro.htm.

CHAPTER 8 ASTEROID KILLERS

180 **soon to be known as "Project Icarus"**: *Project Icarus: MIT Student
Project in Systems Engineering* (MIT Press, 1968).

181 **"Large Asteroid Is Headed for Earth"**: *Lebanon* (Pennsylvania)
Daily News, July 28, 1966; "Icarus Could Be Catastrophic in 1968,"
Poughkeepsie (New York) *Journal*, May 23, 1967.

181 **"Hippies Flee to Colorado as Icarus Nears Earth"**: *New York Times*,
June 14, 1968.

184 **"Students Plot Attack on Asteroid"**: *Des Moines Tribune*, May 24,
1967; "Bomb Asteroid Plan Proposed," *Baltimore Evening Sun*,
May 23, 1967.

184 **disaster film called *Meteor*, stocked with a host of "A-list" and
"A-minus-list" stars**: Fun fact: Natalie Wood, who plays a Russian
interpreter in the movie, was born Natalia Nikolaevna Zakharenko
and actually spoke fluent Russian in the film.

187 **the man believed to have inspired the Dr. Strangelove character
in the Stanley Kubrick film**: On the other hand, Edward Teller was
also an early advocate of taking steps to reduce global warming.

188 **"Killer Asteroid Dooms Earth!"**: *San Jose Mercury News*, *West* magazine, March 22, 1992.

189 **The winner was Sung Wook Paek**: "Paintballs May Deflect an
Incoming Asteroid," *MIT News*, October 26, 2012.

190 an "ion beam shepherd" spacecraft to bombard it with high-speed ions: https://www.popsci.com/whats-best-way-to-protect-earth -from-incoming-asteroids#page-4.

193 the results of an asteroid deflection plan funded by NASA and the National Nuclear Security Administration: Brent W. Barbee et al., "Options and Uncertainties in Planetary Defense: Mission Planning and Vehicle Design for Flexible Response," *Acta Astronautica* 143 (February 2018): 37–61.

195 "There could be cases where the warning time is too short to deflect [the asteroid]": Author interview with Megan Bruck Syal, April 19, 2018. Megan is one of those people who, without in any way meaning to, can make you feel seriously undereducated. She has a BA in astrophysics, a BA in mathematics, a master's of science degree in engineering, another master's in geological sciences, and a PhD in geological sciences from Brown University. It's one example of the kind of brains we have working the asteroid problem.

198 Aerospace engineer Bong Wie doesn't think we should wait: Author interview with Bong Wie, May 3, 2018.

CHAPTER 9 ASTEROID WARS

210 actual scenario used by NASA and FEMA officials: https://cncos .jpl.nasa.gov/pd/cs/ttx14/NASA.FEMA.Exercise.Report.2014.pdf.

212 "the people who joke about aliens or whatever": Author interview with Leviticus A. Lewis, chief of FEMA's National Response Coordination Branch, May 14, 2018.

212 "It was an eye-opener": Author interview with Dan Bout, Assistant Director for ResponseCalifornia, Governor's Office of Emergency Services, May 22, 2018.

215 1977 novel *Lucifer's Hammer*, takes a much darker view: *Lucifer's Hammer*, by Larry Niven and Jerry Pournelle (Del Rey Books, 1977). The "Hammer" is something of a theme for asteroid- and comet-impact books. Arthur C. Clark wrote about an Earth-bound asteroid named Kali in the 1993 sci-fi novel *The Hammer of God*, and in 1983 Wynne Whiteford published *Thor's Hammer*, about an attempt to divert an asteroid to hit Earth.

217 **5th IAA Planetary Defense Conference, hosted by the International Academy of Astronautics in Tokyo in 2017**: https://cneos.jpl.nasa.gov/pd/cs/pdc17/ and https://www.youtube.com/channel/UCpq_pj8aLeFu-oi4Je8_JsQ.

227 **The good news on the Near-Earth Object front**: "National Near-Earth Object Preparedness Strategy and Action Plan: A Report by the Interagency Working Group for Detecting and Mitigating the Impact of Earth-Bound Near-Earth Objects of the National Science & Technology Council," June 2018, https://www.whitehouse.gov/wp-content/uploads/2018/06/National-Near-Earth-Object-Preparedness-Strategy-and-Action-Plan-23-pages-1MB.pdf.

INDEX

2014 AA asteroid, 150
Air Force, 135, 141, 172–75, 206
aliens, 6, 81–82, 108–09, 212, 232, 259n
All-sky Fireball Network, 2
Alvarez, Luis, 108–09, 111, 115–16, 259n
Alvarez, Walter, 105–06, 108–17
   K-T extinction theory of, 109,
      111–12, 117, 118, 125
   popular culture on, 112–13, 114–15
   scientists' criticism of, 113–17, 119
American Association for the
      Advancement of Science, 50, 52
3554 Amun asteroid, 69
Apophis asteroid, 147, 190, 204
Arrhenius, Svante, 64
4581 Asclepius asteroid, 38, 41
Asteroid Day, 169–70, 226
asteroid defense systems, 179–202. See
   also planetary defense
   agency planning meeting on,
      204–05
   comet impacts and, 222–23
   gravity tractor for, 190–91, 199, 227
   Hypervelocity Asteroid Intercept
     Vehicle (HAIV) for, 199–201, 225
   kinetic impactors for, 191–95, 199,
     201, 223
   MIT class project on, 179–80,
     181–84, 185, 190
   NASA workshop on, 186

   nuclear weapons for, 179, 181,
     182–83, 185–89, 195–97, 199, 223
   paintballs for, 189–90
   planning exercises for. See war
     games
   probability data needed for, 197–98
   size of asteroid and, 107
   slow-push methods for, 190–91
   spacecraft flight in, 227
asteroid mining, 66–71
   asteroid water as focus of, 68
   metals and, 68–69
   Meteor Crater and, 54–59, 62,
     64–65, 66, 77
   NASA's interest in, 67–68
   NEOs and, 71
   problems in, 69–71
   space projects in, 66–67
Asteroid Redirect Mission (ARM), 191
Asteroid Terrestrial-impact Last Alert
     System (ATLAS), 137, 155
asteroid threat. See also specific asteroids
   asteroid size and, 157
   asteroids-as-weapons and, 221–22
   comets and, 222–23
   conferences and plans on, 21, 125–26
   defense against. See planetary
     defense
   Defense Department data and, 102
   early lack of interest in, 101, 103

asteroid threat (*cont.*)
  funding of programs for, 158–60
  impact of hit in, 110–11
  Kowalski's research on, 148–51
  Musk's Tesla as, 152
  NASA's goal with, 156, 157–58
  NEOs and, 37–38, 98–99, 156
  number discovered, 156
  officials' communication about,
    213–14
  opinions on, 3–4, 40–41, 176–77
  planning exercises for. *See* war games
  risk corridor for, 205–06, 208, 218,
    219
  search programs and, 134–35
  Shoemaker's surveys of, 97–101
  steady stream of reminder incidents
    on, 101–02
  time frame for planning for, 214
  Torino Scale rating for, 146–47
asteroids, 7–24, 33–44. *See also specific*
    asteroids
  atmosphere as protection against,
    7–8
  cause of glow and fiery tail of, 8
  color and size of, 36
  composition of, 31, 36
  crater evidence of Earth hits by, 44, 74
  creation of, 35
  destructive force of, 10–11
  dinosaurs and. *See* K-T extinction
    theory
  dust from Earth hits by, 12, 101,
    109–10, 118–19, 176, 182, 200
  impact craters on, 37, 97
  interception or deflection of. *See*
    asteroid defense systems
  location in main belt of, 37–38
  mining of. *See* asteroid mining
  "missing planet" search and, 34–35
  moons of, 132
  naming conventions for, 41–43
  nickel-iron composition of, 8–9, 12,
    36, 54, 80
  number of, 35, 41, 43–44
  number of Earth hits over time, 12, 40

  origin of name, 35
  popular culture on, 33–34
  possible collision with Earth. *See*
    asteroid threat
  predicting possible hits by, 39–40
  private ownership and use of, 133
  reported close passes by, 38–39
  search programs for, 132–38, 160–61
  shock wave after impact of, 11–12
  Shoemaker's surveys of, 97–101, 120
  size of, for collision with Earth, 8–9
  spacecraft's photographs of, 37, 132
  speed of, 10–11, 41, 65
  theory on inhabitants of, 87
  unmanned space probes for, 133–34
  water content of, 68
  weaponizing, 221–22
astroblemes, 82, 84, 85, 89. *See also*
    impact craters
Astrogeology Research Program, 95

Baade, Walter, 180
Baldwin, Ralph B., 90
Barringer, Daniel M., 45–48, 54–66, 85,
    93, 231
  business interests of, 46, 62–63, 73
  Coon Mountain claim of, 55–56
  death of, 66, 73
  evidence for impact by, 59–61, 66
  family background of, 45
  financial backers and, 61–62, 64
  first hearing about Meteor Crater
    by, 44, 47–48
  impact theory of, 63–64, 74
  Meteor Crater mining by, 54–59,
    62, 64–65, 66, 73, 77
  mining patent of, 56, 57–58, 252n
  Moon impact crater named for,
    73–74
  Moulton on impact and, 65–66
  personality of, 45–46
Barringer Crater, 73–74
101955 Bennu asteroid, 67, 193–94
Bolden, Charles, 202
bolides, 1–3, 30
Bout, Dan, 212–13

Bowell, Ted, 153
2015 BZ$_{509}$ asteroid, 161

Camargo, Antonio, 118
Canyon Diablo, Arizona, 16, 47, 50–51,
    96. *See also* Meteor Crater,
    Arizona
Catalina Sky Survey, 156, 159, 160
    founding of, 135, 171
    Kowalski's searches for, 141, 143–51,
        154–55
    NEO discoveries by, 151–52, 155, 156
    telescope used in, 141–43, 145–46
catastrophism, 107–08, 114
Center for Near-Earth Object Studies,
    146, 147, 196, 211
Ceres asteroid, 34–35, 37, 42, 187, 200
Chapman, Clark R., 91, 125, 187, 221
Chelyabinsk, Russia, asteroid event
    (2013), 164–70, 202, 212, 226
Chicxulub crater, Mexico, 118–19
Chladni, Ernst, 33
Chodas, Paul, 39, 149, 196, 211, 214,
    217, 221, 224
Christensen, Eric, 156, 157, 159
67P/Churyumov-Gerasimenko comet,
    133–34
coesite, 93, 117
comets 24–29. *See also specific comets*
    comas around, 25–26
    composition of, 24
    fiery tail of, 25–26, 29, 121, 241n
    Great Comets and, 26–28, 241n
    Halley's research on, 27–28
    impact risk of, 222–23
    impending doom and rumors
        associated with, 26–27
    location of regions with, 24–25
    naming of, 28–29
    number of Earth hits over time, 12
    orbits of, 25, 27
    Shoemaker's survey of, 100–01, 120
    speed of, 41
conspiracy theorists, 82, 205, 214, 232
Coon Mountain crater, 49–56. *See also*
    Meteor Crater, Arizona

cosmic pluralism theory, 87
2011 CQ$_1$ asteroid, 151, 163–64
craters. *See* impact craters

2012 DA$_{14}$ asteroid, 163–64, 167–68
Dachille, Frank, 111
Deep Space Industries, 67, 72, 226,
    231, 236
Defense Department (DOD). *See* U.S.
    Department of Defense
Defense Support Program, 102
Deichman, Joe, 179–80, 182, 184
Dekker, Thomas, 23–24, 33, 76
Dietz, Robert S., 89–90
dinosaurs
    Alvarez's K-T theory on, 105–20, 125
    catastrophism belief on sudden
        event affecting, 107–08, 114
    earlier extinction theories on, 111
    extinction events affecting, 258n
    K-T Boundary marking mass
        extinction of, 106
    range of theories on, 106–07, 109
    uniformitarian theory of gradual
        change over time and, 107–08, 114
doomsday drills, 213. *See also* war games
Doppler weather radars, 2
Double Asteroid Redirection Test
    (DART), 192–93, 206, 225
Dreier, Casey, 170
367943 Duende asteroid, 38, 41, 163

Eros asteroid, 133
European Space Agency, 28, 124, 133,
    169, 204, 206, 211, 218, 220, 263n
extinction theories, 111. *See also* K-T
    extinction theory
    catastrophism belief on sudden
        event, 107–08
    uniformitarian theory of gradual
        change over time, 107–08, 114

Federal Emergency Management
    Agency (FEMA). *See* U.S.
    Federal Emergency Management
    Agency

Foote, Albert E., 50, 52, 249n
Franklin, Charles Albert, 48, 247n–48n

Galileo spacecraft, 37, 124, 132
1036 Ganymed asteroid, 98
951 Gaspra asteroid, 37, 132
Gauss, Carl Friedrich, 87–88
Gehrels, Tom, 92, 126, 127, 134, 142,
     172, 174
Geological Survey (USGS). *See* U.S.
     Geological Survey
Gilbert, Grove Karl
     Barringer on theories of, 55, 60
     crater's circular shape and, 85
     influence and reputation of, 52, 60
     Meteor Crater asteroid search by,
          51–54, 65–66
     Moon crater research of, 89
Goldman Sachs, 71
Goldstone Solar System Radar facility,
     Barstow, California, 209,
     262n–63n
gravity tractor, 190–91, 199, 227
Great Comets, 26–28, 241n

Hale-Bopp comet, 27, 28, 262n
Halley, Edmond, 27–28, 111
Halley's Comet, 28, 29, 261n
HAMMER spacecraft, 194–95, 201,
     225
Hatford, England, meteorite shower
     (1628), 23–24, 33
Helin, Eleanor "Glo," 125, 174, 175
     asteroid survey of, 97, 99, 100,
          258n
     search programs and, 134–35
Herschel, Sir John, 88
Herschel, William, 35, 42, 87, 88,
     243–44
Hildebrand, Alan, 118
Hoba meteorite, 31, 242n
Hodges, Ann, 32
Holbrook, Arizona, meteorite fall
     (1912), 14
Holsinger, Samuel J., 47, 55, 56, 59, 62,
     246n–47n

Hug, Gary, 153–54
Hayabusa spacecraft, 134
Hypervelocity Asteroid Intercept
     Vehicle (HAIV), 199–201, 225

1566 Icarus asteroid, 179–85, 190
243 Ida asteroid, 37, 132
impact craters, 19–20, 74, 131
     air travel and discovery of, 85
     alternate theory on, 53, 75, 84, 85,
          86
     asteroids with, 37, 97
     debate over causes of, 86, 89–90, 93
     as evidence of asteroid hits, 44, 74
     K-T dinosaur extinction theory
          with, 117–19
     21 Lutetia asteroid with, 133
     on Moon, 33, 86–90, 95–96, 99,
          132
     number of, 93
     on other planets, 96–97
     Philby's discovery of, 83–84
     planets with, 96–97
     range of locations of, 84–85
     round shape of, 85–86
impact theory
     coesite at Meteor Crater and, 93
     "giggle factor" regarding, 127–28,
          129, 172, 212
     NASA study committee on, 129–30
     scientists' criticism of, 128
     Shoemaker-Levy 9 validation of,
          128–29
     Shoemaker's proof of, 93–94, 97,
          114, 123, 129, 131, 174
International Asteroid Day, 169–70,
     226
International Asteroid Warning
     Network (IAWN), 169, 204,
     217, 227
International Astronomical Union, 135
     Minor Planet Center (MPC) of, 39,
          145, 148–49, 149–50, 171, 201
     number/letter designation assigned
          by, 41–42, 42–43, 73–74
     XF_{11}'s passage near Earth and, 39

iridium, 55, 68, 111–12, 117
25143 Itokawa asteroid, 134

Jenniskens, Peter, 150
Jet Propulsion Laboratory (JPL), 39
    asteroid planning exercise with, 211
    asteroid threats and, 171
    Center for Near-Earth Object
        Studies of, 146, 147, 196, 211
    search programs and, 134, 146
    2008 TC$_3$ impact threat and, 149
Johnson, Lindley, 170–76, 198, 201,
        213, 232
    Air Force background of, 171–74, 175
    first use of "planetary defense" by,
        173–74
    NASA planetary defense office and,
        170–71, 175–76
    planetary defense report of, 173, 174
5905 Johnson asteroid, 175
Juno asteroid, 35, 42
Jupiter, and Shoemaker-Levy 9
        asteroid, 121–30

Kelly, Allan O., 111
Kepler, Johannes, 27, 29
kinetic impactors, 191–95, 199, 201, 223
216 Kleopatra asteroid, 36
Kowalski, Richard, 139–46, 148–53, 231,
        263n
    asteroid threats discovered by,
        148–51, 163
    background of, 152–53, 154
    image monitoring by, 143–48
    telescope used by, 141–43
    workspace of, 142–43, 154–55
7392 Kowalski asteroid, 153
Kring, David, 118
K-T extinction theory, 109–20
    Alvarez's formulation of, 109, 111, 112
    difficulty with concept of, 119–20
    evidence for, 111–12, 117
    impact crater in, 117–19
    K-T Boundary between geologic
        periods and, 106
    popular culture on, 112–13, 114–15

scientists' criticism of, 113–14,
        115–17, 119
    species surviving, 110
    summary of, 109–10
Kuiper, Gerard, 97, 241n
Kuiper Belt, 24, 25, 233, 241n
Kulik, Leonid Alekseyevich, 76–79, 80

2018 LA asteroid, 151
Large Synoptic Survey Telescope, 137
Larson, Steve, 135, 154
Lawrence Livermore National
        Laboratory, 193, 201, 211
Levy, David, 120–21, 124, 125, 129
Lewicki, Chris, 68
Lewis, John S., 66–67, 70
Lewis, Leviticus A., 212, 213–14, 216
Lincoln Near-Earth Asteroid Research
        (LINEAR) project, 135, 151, 174,
        193
Lowell, Percival, 135
21 Lutetia asteroid, 133

M$_{13}$ (star cluster), 136
main belt asteroids, 98, 100
    Jupiter's gravitational pull on, 122
    Kowalski's searches for, 139, 145,
        147, 148
    location of, 37
    number of, 156
    threat from, 37–38
Mars, impact craters on, 96–97
Massachusetts Institute of Technology
        (MIT), 135, 174
    Project Icarus at, 179–80, 181–84,
        185, 190
Matheny, Rose, 151
May, Brian, 169
McLaren, Digby, 111
McMillan, Robert S., 126, 128, 134
Meteor Crater, Arizona, 1–21, 48–54,
        229–31
    Apollo astronauts' visits to, 95
    Barringer's claim on, 55–56
    Barringer's first hearing about,
        47–48

Meteor Crater, Arizona (*cont.*)
    Barringer's impact theory of, 63–64
    Barringer's mining operations at,
        54–59, 62, 64–65, 66, 73, 77
    coesite found at, 93
    destructive force of asteroid in, 10–11
    Foote's report on meteorites at, 50
    Gilbert's search for asteroid at,
        51–54, 65–66
    lack of change or erosion at, 19–20
    local beliefs about origin of, 47
    location and setting of, 13–16, 20
    low hill surrounding, 16–17
    meteorite impact creating, 12, 13, 19
    meteorites in, 16, 17, 20, 48–51, 65, 96
    mining patent on, 56, 57–58, 252n
    Moulton on impact and, 65–66
    naming of, 62, 73
    Native Americans at site of, 48, 49
    private ownership of, 17, 56
    reasons for studying, 20–21
    selling of meteorites from, 50–51
    sheepherders' early claim on, 49–50
    shock wave after impact in, 11–12
    Shoemaker's research on, 91–93, 131
    size of, 17–18, 230–31
    stories about discovery of, 48
    Torino Scale rating for, 146
    tourism and, 17, 63, 73, 233
meteorites, 31–33. *See also* Near-Earth
    Objects; *specific meteorites or sites*
    airburst explosions with, 80–81
    business of selling of, 49
    chances of hitting a populated area
        with, 78–79
    composition of, 31, 48–49
    definition of, 31
    dinosaur extinction from impact of.
        See dinosaurs
    humans hit by, 31–32
    impact creation of, 30–31, 48
    Native Americans' use of, 48, 49
    nickel-iron composition of, 12,
        31, 47, 49, 50, 54, 55, 84, 134,
        242n–43n
    number hitting Earth, 31, 40

    pamphlet (1628) on, 23–24, 33
    size of, 31
    theories on, 33
meteoroids, 29–30
meteors, 29–31
mining. *See also* asteroid mining
    Meteor Crater and, 54–59, 62,
        64–65, 66, 77
Minor Planet Center (MPC), 39, 145,
    148–50, 171, 201
Moon
    impact craters on, 33, 86–90, 99, 132
    men-on-the-moon theory on, 87–89
    meteoric impact causing craters
        on, 96
    meteorites originating on, 33
    NASA's internment of Shoemaker's
        ashes on, 131–32
    space program missions to, 95–96
Morrison, David, 125, 187
Moulton, Forest Ray, 65–66, 85
Multiple Kinetic-Energy Impactor
    Vehicle (MKIV), 201
Musk, Elon, 151–52

National Aeronautics and Space
    Administration (NASA), 164
    asteroid defense systems and, 186,
        191, 192–93, 196, 197, 199
    asteroid mining and, 67–68
    asteroid reports by, 2
    2012 DA_{14}'s close pass and, 164, 168
    FEMA exercises with. *See* war
        games
    funding from, 127, 158–60, 170, 171
    impact study committee of, 129, 130
    lunar internment of Shoemaker's
        ashes and, 131–32
    NEO programs of, 129–30, 133, 141,
        143, 146, 156, 157–58, 160, 170,
        171, 174, 175
    OSIRIS-REx mission of, 67, 194,
        226, 231
    photographs of asteroids and, 37, 132
    planetary defense and, 170–71,
        175–76

planned Psyche asteroid probe of,
36
risk corridor projections of, 205–06
search programs and, 135, 136, 137,
138
Toutatis's passage near Earth and, 40
National Science & Technology
Council, 227
Near-Earth Asteroid Rendezvous
(NEAR) spacecraft, 133
Near-Earth Asteroid Tracking (NEAT)
program, 134–35
Near-Earth Objects (NEOs)
Air Force detection of, 174–75
amateurs' role in detection of,
153–54
defense against. *See* planetary
defense
definition of, 98
description of, 37–38
funding of programs for, 158–60
"giggle factor" on searches for,
127–28, 129, 172, 212
group assignments of, 100
Johnson's interest in, 172–73
Kowalski's searches for, 139–40,
143–51
mining of, 71
naming conventions for, 42–43
NASA's goal in identifying, 156,
157–58
nearness to Earth of, 38
number discovered, 156–57, 170
number of close passes of, 38–39
planning document on, 227
possible collision between Earth
and, 98–99
potential danger of nuclear war
sparked by, 102–03
search programs for, 126–27, 133,
134–38, 155–56
Shoemaker's surveys of, 99–101
Spaceguard searches for, 126,
129–30, 141, 261n
Spacewatch sky survey of, 126, 127,
134, 142, 155, 172

Torino Scale rating for, 146–47
warning network for, 169
Near Earth Objects Dynamic Site
(NEODyS), 146
NEAR Shoemaker mission, 133
NEOCam (Near-Earth Object
Camera), 138, 158, 225–26
NEOs. *See* Near-Earth Objects
NEOWISE, 137–38, 155
New York Times, 61, 73, 78, 114–15, 116,
124, 129, 181, 184, 187
Nördlingen, Germany, impact crater,
93
nuclear weapons, 179, 181, 182–83,
185–89, 195–97, 199, 223
Nye, Bill "The Science Guy," 99, 169

Oort Cloud, 24, 25, 233, 241n
OSIRIS-REx spacecraft, 67, 194, 226,
231
'Oumuamua asteroid, 160
Outer Space Treaty (1967), 71, 188, 196

Paek, Sung Wook, 189, 190
Pallas asteroid, 34–35, 42
Palomar Asteroid and Comet Survey,
100–01, 120, 127
Palomar Planet-Crossing Asteroid
Survey, 97–98, 99
Panoramic Survey Telescope and
Rapid Response System (Pan-
STARRS), 137, 155–56, 160, 225
Park, Robert L., 187–88
Patterson, Clair, 96
Penfield, Glen, 118
Pew Research, 72, 160
3200 Phaethon asteroid, 136
Philby, Harry St. John B., 83–84
Piazzi, Giuseppe, 34, 37
Pigott-LINEAR-Kowalski comet, 151
planetary defense, 170–76. *See also*
asteroid defense systems
current status of, 224–28
first use of phrase, 173–74
Johnson's report on, 173, 174
NASA's office for, 170–71, 175–76

planetary defense (*cont.*)
  planning exercises for. *See* war
    games
  when to start work on 198–99
Planetary Defense Conferences, 21, 217,
    222, 223, 224
Planetary Resources, 67, 68, 72, 226,
    231
Planetary Society, 5, 99, 154, 170, 228
platinum, 55, 56, 68–69, 70
Pluto, 4, 135
Project Icarus, 179–80, 181–84, 185, 190
16 Psyche asteroid, 36, 41

radar, 2, 3, 136, 141, 172, 209, 262n–63n
2206 RH$_{120}$ asteroid, 151
Richardson, Robert S., 180
risk corridor, 205–06, 208, 218, 219
Robertson Panel, 108–09, 259n
Roosevelt, Teddy, 46, 62
Rosetta space probe, 133–34
162173 Ryugu asteroid, 134

Sagan, Carl, 99, 188
Sandlot Observatory, 153–54
Sandorff, Paul, 180, 181, 182, 184
satellite moons, 36, 132
satellites
  asteroid reporting using, 2, 102, 124
  commercial asteroid exploration
    with, 67
  first launch of, by Soviets, 94
  NEOs tracked by, 172, 174–75, 226
  search programs, 132–38
Shaddad, Muawia, 150
Shoemaker, Carolyn, 91, 131
  comet discovery by, 120–21
  comet survey and, 100–01, 120, 127
Shoemaker, Eugene "Gene," 90–94,
    97–103, 120, 133
  Alvarez's K-T extinction theory and,
    109
  background of, 91–92
  death of, 130–31
  impact theory proof of, 93–94, 97,
    114, 123, 129, 131, 174

lunar internment of ashes of, 131–32
lunar missions and, 95
Meteor Crater research of, 91–93,
    131
NASA impact study and, 129
Palomar surveys of, 97–98, 99,
    100–01, 120, 127
personality of, 90–91
Shoemaker-Levy 9 discovery and,
    121
Shoemaker-Levy 9 comet, 120–30
  commotion over impending
    collision with Jupiter, 123–24
  discovery of, 120–21
  effect of collision with Jupiter,
    124–30
  impact theory validated by, 128–29,
    174
  Jupiter orbit of, 121–23
space exploration programs, 94–96,
    159–60, 171, 179–80
Spaceguard, 126, 129–30, 141, 261n
Space Mission Planning Advisory
    Group (SMPAG), 169, 204, 218
Spacewatch sky survey, 126, 127, 134,
    142, 155, 172
Sputnik satellite, 94
spy satellites, 2, 102, 174–75
2867 Šteins asteroid, 133
Strategic Defense Initiative (SDI),
    186–87, 187–88
Syal, Megan Bruck, 195, 196, 197
Sylacauga, Alabama, bolide (1954), 31

2015 TB$_{145}$ asteroid, 37
2008 TC$_3$ asteroid, 148–50
Teller, Edward, 187, 188
Tempel 1 comet, 192
Time (magazine), 113, 114, 124, 184
Torino Scale, 146–47, 190, 263n
4179 Toutatis asteroid, 39–40, 42, 263n
2014 TTX asteroid scenario, 203–10
Tunguska meteorite, Siberia (1908),
    75–82, 157
  alien and other alternative theories
    on, 81–82, 254n

Asteroid Day on date of, 169–70
cause of, 79–81
description of event in, 75–76
impact's devastation in, 77–78, 80
Kulik's theory on cause of, 77 79
lack of meteorites after, 78, 79, 80
later scientific expeditions to, 79
media reaction to, 78–79
power of explosion in, 80, 81
Torino Scale rating for, 146

unidentified flying objects (UFOs),
    5–6, 27, 82, 108–09, 259n
uniformitarian doctrine, 107–08, 114
United Nations (UN), 164, 169, 170,
    199, 218
U.S. Air Force, 135, 141, 172–75, 206
U.S. Department of Defense (DOD),
    102, 158, 180, 204
U.S. Federal Emergency Management
    Agency (FEMA)
disaster planning by, 203–04, 227
NASA exercises with. *See* war games
U.S. Geological Survey (USGS), 51, 52,
    89, 91, 95

Vesta asteroid, 35
volcanos

Earth craters caused by, 75, 84, 85
Moon craters caused by, 86–87, 89,
    90
Volz, Fred W., 50–51, 63, 249n–50n
von Gruithuisen, Franz, 88
Vredefort crater, South Africa, 85, 93

Wabar, Saudi Arabia, 83–84
war games, 203–13
    international exercise in, 217–21
    officials' reactions to, 212–13
    purpose of, 210–11, 223–24
    small asteroids used in, 211–12
    2014 TTX planning exercise in,
        203–10
Weissman, Paul, 125
Wellerstein, Alex, 185–86, 189
Wie, Bong, 198–99, 200, 201
Wolfe Creek crater, Australia, 85
Worden, Pete, 102–03
$WT_{1190}F$ space junk, 150

1997 XF_{11} asteroid, 39, 244n

Yarkovsky Effect, 190, 194, 232
Yeomans, Donald, 39

Zahnle, Kevin, 128